Introducing
Radio Control
MODEL
BOATS
Vic Smeed

Introducing
Radio Control
MODEL
BOATS
Vic Smeed

ARGUS BOOKS

ARGUS BOOKS LTD
Wolsey House
Wolsey Road
Hemel Hempstead
Herts HP2 4SS

First published in 1983
Reprinted 1984, 1986 (Twice), 1987
© Vic Smeed

ISBN 0 85242 803 0

Set by SIOS Ltd, 111–115 Salusbury Road
London NW6 6RL
Printed and bound by Adlard & Son Ltd, Pixmore Avenue,
Letchworth, Herts SG6 1JS

Contents

1   The Radio   7

2   The Extra Controls   19

3   The Boats and Competition Classes   29

4   The Construction   39

5   The Power   49

6   The Installation   63

7   Checks and Operation   79

8   Maintenance and Fault-finding   85

Modern boating includes multi-racing. Here four boats are covering the end leg of the course at 40 m.p.h. plus, avoiding a fifth which has stopped.

At the other extreme are scale models of remarkable complexity, such as this West German model photographed at a European Championship.

CHAPTER ONE

The Radio

Radio control of models goes back over fifty years, but practical equipment for the average modeller to buy only began to be available from 1950 onward, and, compared with present-day commercial sets, it was pretty crude and limited in its application. In those days, most people hadn't heard of transistors and valves (tubes) were used in both transmitter and receiver. This meant a separate filament-heating battery and a high-voltage one for operation; a typical transmitter required a 1½v heater and a 120 or 135v output battery, while 1½v and 67½v supplies were quite normal for receivers. And only one model could be operated at a time, except for isolated instances where equipment had been hand-built by radio experts.

Development of radio control equipment

The earliest model equipment, in the 1920s, used a spark transmitter, the spark generated by an induction coil operating a coherer receiver. An electrical spark releases a surge of energy which oscillates between inductance and capacitance, dying away quickly but producing a momentary radio emission over a broad band of frequencies, which is why extraneous electrical sparks cause radio interference. A 'coherer' receiver included a shallow tray of iron filings and on receipt of the brief signal the filings formed magnetically into a line connecting two contacts, allowing a current to flow to a control mechanism. To stop the current flowing, the filings tray had to be tapped, and an automatic mechanism tapping the tray every second or two was incorporated!

Early commercial radio gear was similar in that it simply detected the presence or absence of a carrier wave; the receiver circuit was tuned to be on the point of non-oscillation, and the received carrier wave prevented oscillation. The transmitter had only a press button which, when operated, stopped the carrier wave, and the receiver detected the current change as the circuit went into oscillation. A relay operated by the change in current closed contacts through which a circuit was made, and the degree of control depended on the ingenuity with which this circuit was used. Normally it operated a sequential escapement mechanism.

Greater reliability was achieved by introducing a modulation (tone) stage, the carrier wave remaining on and the switching of the relay effected by the presence or absence of the tone. By introducing a number of tones by means of vibrating metal reeds of different lengths and hence different frequencies, matching reeds in the receiver could be selected to give several switching circuits. Escapements developed into servos of two types, either self-centring or progressive. The former moved to full-out on receipt of a signal, but returned automatically to centre when the signal ceased; to hold an intermediate position meant 'pulsing' the transmitter key manually. A progressive servo moved out on signal, but stopped at the position reached on cessation and needed a second, different signal to move back in the other direction.

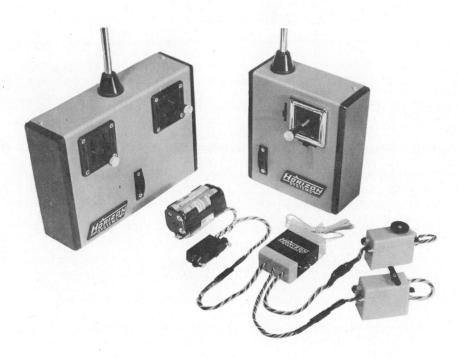

1·1 Two functions on one transmitter stick or separate sticks are alternatives. Battery pack, switch, receiver and two servos comprise the on-board equipment.

Transistors could do the same things with greater stability and reliability, in less space and with a much lower single current supply. Different frequencies could be provided electronically, and switching could be effected by transistors rather than relays, although reeds and relays remained in use for some years. However, all but a few hand-built sets were still of the super-regenerative type which allowed only one model to operate at a time, and it was not until the early 1960s that transistor prices and development permitted commercially-priced superheterodyne equipment to be produced.

Superheterodyning basically means mixing a signal frequency generated within the receiver itself with the aerial signal, the output from the mixer being four frequencies – the incoming, the self-generated, and the sum of and difference between the first two – which are well separated so that it is possible to select the last and amplify only that. A very small difference in the transmitted signal would produce a much larger difference in the last signal (called the intermediate frequency or I.F.) which would not be accepted by the selective I.F. amplification stages. Originally, a separation of 50 kilocycles (now kiloHertz or kHz) was required between transmitted signals, thus giving six 'spot' frequencies in the 27MHz model control band, which actually extends from 26.96 to 27.28MHz, a spread of 320kHz. Later equipment, with greater selectivity, allows approximately 25kHz separation, giving 13 usable frequencies, while the newest UHF system drops to 10kHz and thus allows 32 modules using the newest equipment to operate simultaneously. Difficulty may arise when someone using older equipment may wish to run, since the lack of selectivity could mean that other sets interfere.

Modern radio control equipment

Modern radio gear works on a digital output system where the transmitted signal is divided into constant-length frames, perhaps sixty a second being sent, each frame containing information pulses and a synchronising or resetting pulse. A two-function set will carry two information pulses and a synchroniser; the length of the information pulses is varied by moving the two controls on the Tx. When the Rx is switched on, it drops into step with the synchronising pulse by means of its decoding section, which also identifies the information pulses and routes them through to their appropriate control circuits, which include the servo mechanisms which actually produce the physical movements which may be required.

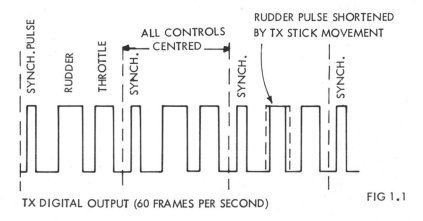

FIG 1.1

TX DIGITAL OUTPUT (60 FRAMES PER SECOND)

9

The circuitry in the servo consists of an amplifier which can switch the servo drive motor in either direction, and a feedback potentiometer driven by the motor and related to the position of the servo output arm. The amplifier and pot are in balance when at rest, which is when the amplifier is satisified that the position of the output arm corresponds to the length of the information pulses coming in. If the pulse length changes, as happens when the operator moves the relevant Tx control, the amplifier will switch on the motor moving the output arm until the feedback through the pot is once again in balance and the motor is switched off.

Radio signals are electromagnetic waves created by oscillation at high frequency of an electric current in a transmitter and radiated through an aerial or antenna. The waves from a simple aerial radiate concentrically, like still-water ripples from a rising fish, and travel at the speed of light, decreasing in intensity but maintaining the same frequency, and having the property of passing through or round most objects in their path. Everything is subject to these waves, but if another aerial is arranged to receive them, and connected to a receiver with selective circuitry, the receiver can reproduce the output of the transmitter to which it is matched.

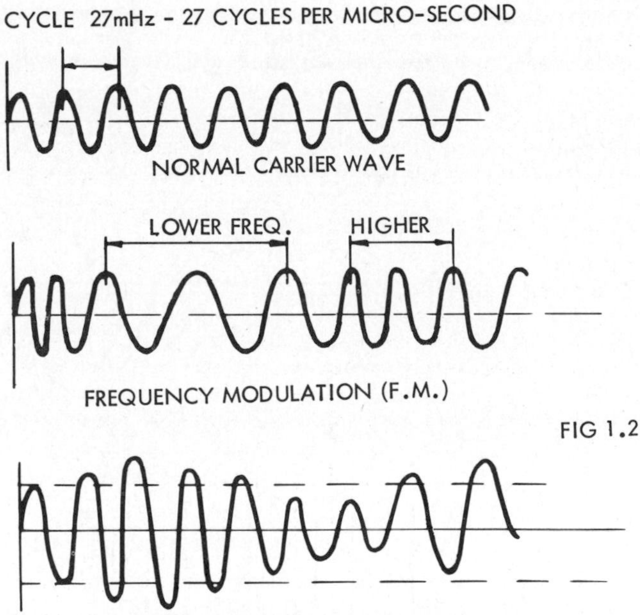

CYCLE 27mHz – 27 CYCLES PER MICRO-SECOND

NORMAL CARRIER WAVE

LOWER FREQ.   HIGHER

FREQUENCY MODULATION (F.M.)

FIG 1.2

AMPLITUDE MODULATION (A.M.)

The transmitter output in our equipment consists of the basic radio wave created by a particular frequency of oscillation which, before radiation, is modulated by slight variations in its frequency (frequency modulation, or FM) or by variations in its amplitude, or strength (amplitude modulation, or AM). This modulation of the basic carrier wave is, effectively, introducing a coded signal which the receiver detects and separates.

The precision of transmitted and accepted signals is achieved by the use of 'crystals' which are fine slices of quartz ground to resonate on very exact frequencies and introduced into the Tx (transmitter) and Rx (receiver) circuits. Because it is mixed with another signal, the Rx crystal is actually different from that for the Tx, and it is thus important when plugging in a different pair of crystals that each is plugged into the right part of the equipment. Each crystal is mounted in a small canister fitted with two pins and is identified by the letters 'Tx' or 'Rx', plus the frequency and a colour coding.

Crystals seem quite solid little components, but the thin slice of quartz inside the case is quite delicate and can be damaged by rattling about in a tool-box. Keep them in a separate tin or box with a little padding, keep them clean and avoid bending the pins and they will last for years. One other small point – crystals from different manufacturers may vary, and you should always use the crystals supplied for, and matched to, your particular make of equipment.

The Transmitter and Receiver

A transmitter will have control sticks, trims, an on/off switch, a telescopic aerial, and probably a small meter showing the state of charge of the batteries. It may have a charging socket and access for crystal changing, and in some cases what is called a 'buddy box socket', which allows a lead from another Tx to be plugged in for instructional purposes; this is mostly used when learning to fly aircraft. Some Txs are fitted with an eye for a neck strap, though this is not really necessary with two-function equipment.

The Rx is even simpler, just a small plastic box with a flexible wire aerial emerging, sockets for the servo wires and a battery input lead, and a crystal, usually fairly prominent for easy changing. A battery pack is separate, and the harness for this usually includes the on/off switch and possibly a charging socket. The usual on-board installation thus consists of four items – Rx, battery pack and two servos.

Most transmitters use lever-type control sticks, that for the rudder moving left or right from a lightly sprung centre position. Moving the stick to one side will cause the information pulse to be lengthened, to the other side shortened. The transmitted signal is thus changed and, after filtering through the Rx decoder, is 'read' by the servo amplifier, which instantly switches on the servo motor to drive the output arm to the position at which equilibrium is regained. If the Tx control is moved halfway to the left, the servo arm will respond by moving halfway to the left; the servo movement is always proportional to the degree of control stick movement. A tiny movement of the stick should produce an equivalent tiny movement at the servo, and the exactitude with which the servo stops (its 'resolution') should be within about one degree.

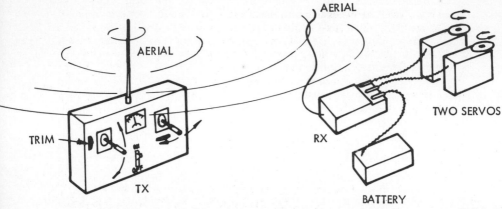

FIG 1.3

Each control stick has a small rotary adjustment set parallel with the control movement which is a trimmer having perhaps 10% of the total movement, and this is used to institute a bias making operation more convenient. For example, if a boat does not run quite straight, the rudder trim can be used to make it do so, without affecting the neutral sprung position of the main control. The trim can also be used for very fine control, perhaps for engine tick-over or very gentle turns. Permanent requirement of a trim bias is an indication that the model itself should be adjusted. One point about trim adjusters which is not always appreciated is that application of full trim and full stick gives a slightly greater control movement than is available from stick alone. This can be useful to, say, shut off an engine or switch something on, since if the trim lever is left central, full stick can be applied without operating the shut-off or switch; to operate it, the trim is pushed hard over and the device will be triggered or switched as soon as full stick in the same direction is applied momentarily.

Two-function systems

Almost all boats use two-function radio, the sprung side-to-side rudder stick mentioned and a second Tx stick at 90°, i.e. up and down, which is not usually sprung but operates against a light ratchet. This second control is used for throttle on an i.c. engine, speed control (with, possibly, ahead or astern) on an electric model, or sails on a yacht. Controls should always be naturally sensed, i.e. to go faster or let sails out, the stick is pushed forward (up). A slight snag for beginners is that left stick for left rudder is easy when the model is travelling away from the operator, but a turn to the operator's left when the boat is heading towards him requires right rudder. Initially this is best handled by thinking only of the boat and which way it must turn, left or right, forgetting about one's own left or right. It soon becomes automatic, but it is an interesting aside that the problem does not seem to arise with the one or two wheel-steering transmitters available; with these, the wheel is just turned whichever way the boat is required to turn.

Six-function systems

More controls are unnecessary with the average boat but sets with six or more functions are available. A six-function set will send six information pulses and a synchronisation pulse in each signal frame, of course, but the noticeable difference will be that the two Tx sticks each have one side-to-side function and one up-and-down; in addition there will be two ratcheted rotary or slide switch-type controls. Uses for additional functions might be independent control of two or more motors, needle-valve adjustment, trim tab angle, and novelty effects such as fire monitors, lifeboat lowering, siren, radar and so on.

1·2 Modern, inexpensive two-function dry battery equipment. Components to be carried in the model are exactly as in the previous photograph.

Batteries

There is a choice open to purchasers of radio equipment – whether to pay a considerably lower initial price and use dry batteries, or to pay the extra from the outset to cover rechargeable cells and the necessary charger. In the first case running costs will be noticeable, but the operating costs with rechargeable cells are negligible.

Dry batteries, as sold for flashlights, etc. are non-rechargeable and, deliver, initially, 1.5 volts per cell, reducing slowly with use. There are high power (HP) versions which, though costing a little more, do appear to give about twice the useful life and are therefore well worth the extra, and it is also advisable to use leakproof cells, unless you make a habit of removing all batteries from the equipment between outings. Alkaline cells last even longer, but the cost of a set of

13

these is well on the way to that for rechargeable cells.

Usual voltages are 12v for the Tx and 6v for the Rx, i.e. eight dry batteries for the Tx and four for the Rx, but the components used in modern sets have a very wide tolerance and replacing the dry cells with 1.2v rechargeable ones (giving 9.6v for the Tx and 4.8v for the Rx) will normally result in no noticeable difference in average operation. Current drain on the Tx is reasonably constant and quite light, and a set of HP dry cells could last for a season of average use; the meter showing battery condition is a useful check, but more important is a physical inspection to ensure that there is no corrosion or other sign of cell damage.

In the case of the Rx, current drain can be much heavier, even with two lightly-loaded servos used for rudder and speed control, and a battery life of 2½ hours or so is all that can be expected. Falling battery output reduces range, permits greater interference effects and may make servo response erratic. It is necessary to be ruthless with dry cells – if there is the slightest doubt, throw them away.

Clearly, rechargeable cells should certainly be considered for the Rx; at current prices they can cover their cost, and that of the charger, in about three months of operating once a week. Once the charger is acquired, it becomes convenient to charge the Tx too and avoid the need to buy dry cells altogether!

The rechargeable cells used are of the nickel cadmium type, generally known as nicads, and they can be obtained in physical sizes identical to dry cells, or as 'button' cells, usually known as Deacs (after the German D.E.A.C. firm which first introduced them for model use). To match voltages exactly, ten are needed

1·3 The white-cased cells are 1·2v 1·2amp/hour nicads, compared with a standard dry battery.

14

for the Tx and five for the Rx, since they deliver only 1.2v each; however, their discharge characteristics are such that they deliver a constant voltage for almost their full specified time, falling away quite rapidly at the end. In fact, a nicad measuring 1.1v is effectively fully discharged. Most modern R/C equipment is designed to work from a range of voltages and, as mentioned, in practice little operational difference is likely to be discerned if dry cells are replaced by nicads of the same physical size.

Many sets are sold with nicads already in place, and in transmitters, these are usually button cells stacked and enclosed in a thick plastic sleeve shrunk on during manufacture. A charging socket is built into the Tx case and a charger is supplied having the correct connections and charging rate. Receiver nicads tend to be built into a rectangular case, most often of the same dimensions as the frame commonly used to hold four dry cells in many types of electronic equipment; this is supplied with a wiring harness bearing an on/off switch and a charging socket, the harness plugging into the Rx. The charger supplied has a second output fitting the Rx battery socket and with the appropriate charging rate automatically supplied. Because the Tx and Rx voltages are different, the two charger outputs have different connections so that incorrect coupling cannot occur.

Where nicads have been substituted for dry cells, a maker's charger may not be available, but there are many chargers commercially available which are entirely suitable once the correct charge rate has been established. Most radio equipment

1·4 Four dry batteries in a typical holder for radio equipment (left) compared with an equivalent sealed nicad pack.

uses cell blocks with a capacity of 500 milliamp/hours, i.e. they will deliver a current of 500 ma (½ amp) for, nominally, one hour before flattening. Recommended charge rate for modern cells is 10% of capacity for ten hours, theoretically, but in practice it is advisable to allow ⅓ extra, or a little more, to offset the degree of inefficiency which is inherent in the charging procedure. Thus the standard charge for 500ma cells is fourteen hours at 50ma.

If heavy duty Rx batteries are used (say to drive a sail winch as well as the Rx and rudder servo) of, typically, 1.2a/h capacity (1200ma/h) the charge rate would be 120ma for fourteen hours. Current belief is that lower (trickle) charge rates may actually reduce cell capacity, and faster charges are inadvisable for similar reasons, plus shortening cell life, unless the cells are of a special fast-charge type. Normal advice is to charge for fourteen hours irrespective of whether the cells are only partially discharged, or if the cells have been allowed to flatten completely, a twenty-four hour charge should be given. Exceeding the charge time is unlikely to harm the cells, and it is in fact better to charge for too long than too short a period.

Certain legends have grown up around nicads, but only one has any substance. If a stack of cells is allowed to go completely flat, one or more cells will reach 0v before the others and may thus take a tiny charge from the others at reverse polarity. Normal charging will rectify this, but if there is any doubt, two or three cycles of full charge followed by full discharge (at about 200-250ma) and full charge again will restore the cells to normal. There is, in fact, little that can be done to abuse nicads (other than incorrect charging) which cannot be corrected by two or three charge/discharge cycles. The only attention they need is to be kept clean and to have the soldered wires of the connection tags checked periodically.

If cells are to be joined into a pack, use heavy gauge wire bus-bars, scrape and tin the cells and wires, and join quickly using a large, hot iron (60-80 watts or more) applied for the shortest time commensurate with a sound joint.

Fast-charge nicads are of a special construction, notably in that they are vented, and can be charged in a matter of fifteen to twenty minutes or so at very high currents. They can be used in radio equipment, but if so are usually treated as normal nicads, i.e. charged at the 10-hour rate. Care is needed in fast-charging, and manufacturers' instructions should be followed closely.

Range

A common question is the range available with model radio, and under normal circumstances it is likely to be about 1½ miles, or, at ground level, allowing for gradual weakening of the signal by ground or water absorption, not less than ½ mile. Boats are seldom operated at greater ranges than 200-250 yards, as beyond that it is increasingly difficult to see what the model is doing.

Colour coding

Colour coding is a quick and convenient means of identification and control. The original six frequencies were each given a solid colour, and when the 25MHz intermediate frequencies appeared they were given the adjacent colour each side, since they 'split' the gap between those frequencies; they are referred to as 'splits'. Each transmitter is required to show a pennant or ribbon(s) on its aerial so that other users can tell at a glance what frequencies are in use, and they can then select a set of crystals for a frequency not in use. Competition and club activities are usually governed by a frequency board providing coloured clothes pegs or similar tokens, and only the current holder of a particular peg may switch on his equipment.

The colours of frequencies are:

Colour	Tx Frequency	Rx Frequency
Black	26.975	26.520
Brown	26.995	26.540
Brown/Red	27.025	26.570
Red	27.045	26.590
Red/Orange	27.075	26.620
Orange	27.095	26.640
Orange/Yellow	27.125	26.670
Yellow	27.145	26.690
Yellow/Green	27.175	26.720
Green	27.195	26.740
Green/Blue	27.225	26.770
Blue	27.255	26.800
Purple or Blue/Grey	27.275	26.820

These colours and frequencies are common to most countries, but in some, additional wavebands are allowed. The 35MHz band carries a yellow and white check flag, 40MHz green and white check, 53MHz a black flag, 72MHz a white one, each in addition to a solid colour indicating the specific frequency.

1·5 Transmitter and receiver crystals with one pair in position in the equipment.

17

Fast electric installation with short rudder linkage and second servo operating an on/off microswitch. Note water cooling pick-up, often needed with electrics.

Another fast electric, this one with rudder servo alongside rudder and two-speed switching using two microswitches and a sprung arm to absorb excess travel.

18

CHAPTER TWO

The Extra Controls

Most radio starts life as suitable for aircraft, so that a two-function set will normally have two conventional servos originally intended for rudder and elevator, or rudder and throttle, on a flying model. For an internal combustion powered model boat, such servos are entirely suitable, one for rudder and one for throttle, except when very heavy loads are applied, such as might occur with an unbalanced rudder in a fierce propeller stream on a large R/C hydroplane. Even here there are 'heavy duty' servos made for some aircraft purposes, although it is not uncommon in the U.S.A. for two normal servos of a three-function set to be linked together to operate one rudder. Three-function equipment can also be used if twin motors with independent throttles are fitted.

Conventional servos, however, encounter limitations when the control of electric power or sails is required, or when several operations controlled by one channel are desirable. In such cases an alternative mechanism is required, or the servo may be adapted sometimes to switch or initiate actions by a secondary mechanism. Since this book is introductory, modifications to the radio circuitry are outside its scope, and it is proposed, therefore, to mention only commercial devices which are intended to be plugged into a Rx in place of a normal servo, or the use of a servo to operate switching systems.

Most servos have a rotary output and are provided with a disc which fits on to a squared or splined output shaft, retained by a small screw. Some equipment uses a double-ended arm in lieu of a disc, some uses a final gear to drive a linear rack, so that a tag protrudes from the servo case and slides back and forth along a slot. This last has the advantage of a simple linear motion with no sideways displacement of the push-rod, at least at the servo end. In all cases the typical movement available on the push-rod is about ⅜in. (9mm) each side of neutral; this cannot be increased on a linear servo, but alternative output discs or arms giving greater throw are available for many rotary output types, or extensions can be fitted to standard discs. Spare discs are not difficult or expensive to obtain for the majority of popular sets.

Extension arms

Extending a rotary servo arm will provide a greater amount of travel at the arm

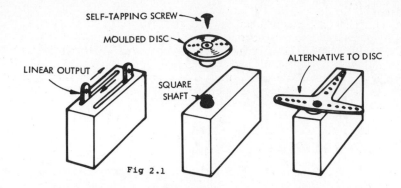

SELF-TAPPING SCREW

MOULDED DISC

LINEAR OUTPUT

ALTERNATIVE TO DISC

SQUARE
SHAFT

Fig 2.1

tip, but the forces resisting travel will be applied through a greater moment, i.e. a mechanical disadvantage is being built in which may result in heavier current consumption, damage to the gear output train, or even a burned-out servo motor. Unless, therefore, the loads are known to be light, extensions are not always a good idea. Where one might be useful is in connecting a rudder with a long tiller; if the servo has a 2in. arm and the rudder a 2in. tiller, the degree of movement and the loads will be the same as a ½in. arm and a ½in. tiller, but the amount of 'slop',

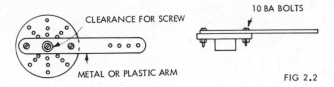

CLEARANCE FOR SCREW

10 BA BOLTS

METAL OR PLASTIC ARM

FIG 2.2

i.e. free movement because of necessary clearances in the arm and tiller holes, will be proportionately much less, so that control of the rudder will be more precise.

Another use for an extension arm is to carry a lightly spring-loaded electrical wiper over a printed circuit switching plate. The greater the radius swept by the wiper, the easier it is to select separate contact areas on the printed circuit board; at ½in. radius the contact will travel about ¾in. for a 90° arc, at 1½in. radius, 2¼in. or thereabouts, so that contact segments can be three times as wide (3.14 times, to be exact).

SPRINGY WIPER

PRINTED CIRCUIT
BOARD

FIG 2.3

Microswitches

Perhaps the most common modification to a servo disc is to shape it into a cam, or bolt a shaped cam to it, to operate microswitches mounted close to the servo. One microswitch each side of the neutral position is usual, but there is no reason why two or more should not be used each side, staggered to operate sequentially. There have been mounts made for this purpose commercially, but it is a simple matter to make a ply shelf with a cut-out for the servo body and sufficient space to accommodate the switches each side.

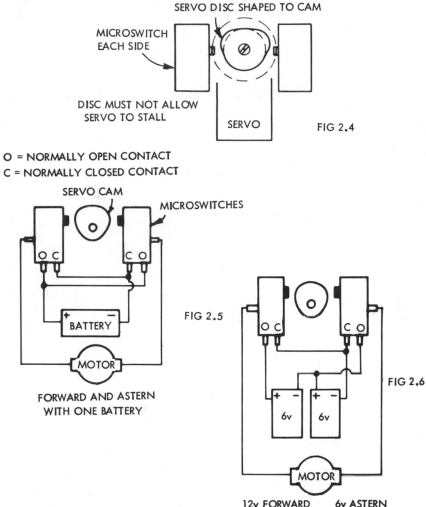

SERVO DISC SHAPED TO CAM

MICROSWITCH EACH SIDE

DISC MUST NOT ALLOW SERVO TO STALL

SERVO

FIG 2.4

O = NORMALLY OPEN CONTACT
C = NORMALLY CLOSED CONTACT

SERVO CAM

MICROSWITCHES

BATTERY

MOTOR

FIG 2.5

FORWARD AND ASTERN
WITH ONE BATTERY

6v   6v

FIG 2.6

MOTOR

12v FORWARD   6v ASTERN
(BATTERIES CAN BE VARIED)

21

Uses for switched circuits of this type are fairly obvious, the most frequent being stop, half-speed, and full-speed on the main drive motor in an electric boat. Or it could be ahead and astern, with stop (or slow ahead with a third, central switch) when the servo is centred. Whenever half-speed (or, more correctly, half-voltage) is used, with a pair of batteries in parallel for half-speed and in series for full-speed, it is always desirable and sometimes important to have two identical batteries to avoid current flowing between them and causing possible damage. For ease of wiring, a double contact, or two simultaneously operated switches would be desirable in such an arrangement, the second contacts connecting the 'middle' positive and negative of the batteries to put them in series for full-speed.

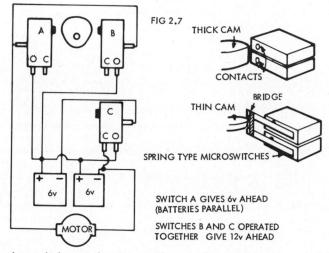

FIG 2.7

THICK CAM

CONTACTS

THIN CAM   BRIDGE

SPRING TYPE MICROSWITCHES

SWITCH A GIVES 6v AHEAD (BATTERIES PARALLEL)

SWITCHES B AND C OPERATED TOGETHER GIVE 12v AHEAD

Two microswitches can be operated by a single cam if they are 'sandwiched' one above the other and either the cam is thick enough to cover both buttons or the buttons are bridged so that a thin cam bears on the centre of the bridge. It really depends on the form of microswitch.

It is worth mentioning that microswitches are akin to the relays which used to form part of radio equipment, so that some of the clever relay-operated mechanisms detailed in books and magazines of the 1955-70 period can easily be adapted. As an example, an impulse selector (a ratchet and pawl wheel with many sets of contacts) can be operated by quick blips of a microswitch, to give perhaps a dozen or more sequential controls from a single switch; the 'other side' of the servo could be used to operate a second switch returning the selector to a standard position. Thus the operator would know that eight blips always brought into operation, say, the siren and ten blips lowered an anchor, etc.

Toggle switches

An alternative to microswitches is the type of three-position (centre off) toggle switch used on car dashboards; these can be found in miniaturised form and a wire push-rod from a conventional servo disc can be passed through a hole drilled in the toggle, with a nut or soldered washer each side of the toggle. More than one switch can be operated simultaneously by one servo, provided the spring action of the switches is not excessively strong.

The advantage with physical switching of this type is that it is easily understood by modellers without electronic knowledge. Whether wipers or switches are used, the result is simply to close electrical circuits, and what use is made of the circuits is up to the ingenuity of the individual.

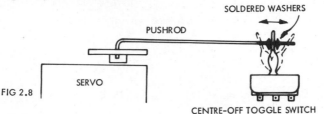

FIG 2.8

SOLDERED WASHERS

PUSHROD

SERVO

CENTRE-OFF TOGGLE SWITCH

Speed controls

Electronic enthusiasts are more likely, for basic speed controls, to make up a transistorised unit, either from one of several circuits published in model magazines or from one of a small number of kits available. Most of these plug into the receiver in place of a servo; several have been developed primarily for electric car racing, but apply equally to boats, except that some provide only forward speeds. Most of the car ones are sold as finished units, sometimes providing receiver current supply as well, and prices range from about the cost of one normal servo up to the equivalent of three servos. Additionally there are speed controllers (giving infinitely variable ahead and astern with a stop position) which are operated by a push-rod from a standard servo; these too are available as completed units, very simple to wire in, and costing around the equivalent of one and a half to two conventional servos.

For the average electric boat modeller, such units are preferable to microswitch systems unless something basically very simple, such as on/off or ahead/astern, or very lightweight, is required. As with the radio itself, once the initial cost is met, a long life can be expected, and trouble-free installation is easy and quick. There are three points which should, however, be considered: first, the capacity of the unit must be adequate for the boat's power, second, if top speed is important, it should have a full-speed by-pass, and third, make sure that it is not forward-speed only if astern manoeuvring is required.

The first point is fairly obvious, but since the cost usually increases with the capacity, there is little point in paying more than is necessary; additionally, there could be current losses at the low end of the unit, reducing running time. Speed controllers usually carry a label giving maximum capacity, either in volts and amps or watts (= volts × amps). If a boat's motor draws a maximum of 3 amps at 12 volts (36 watts), there is little point in fitting a controller with a capacity of, say, 200 watts (16½a at 12v or 10a at 20v). For the average scale ship type of up to about 4 ft. (1200mm) a 60 watt controller should be adequate; 200 watts would only be required with a fast scale launch, a big scale ship, or a speed model of one of the smaller classes.

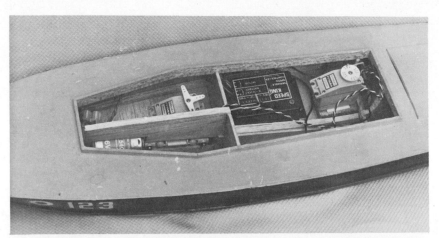

2·1 *Speed control in an electric boat. The speed controller, centre, is operated by the servo on the left.*

2·2 *Close-up of the speed controller above, showing the simplicity of the wiring. A conventional servo moves the sliding control arm.*

Top speed is only obtainable when the controller has a by-pass or a microswitch operated at the full-speed position. This is because most transistorised controllers work on a pulse system and the maximum pulsed output is perhaps only 95% of the battery's output capability. By-passing the controller allows full battery power to be supplied to the motor, and many controllers now have this facility built in.

Mention is made of reversible potential (point three) only because several of the car speed controllers do not cater for reversing, so however good one might be, if it won't allow a boat to go astern it may not be the ideal choice. Speed models rarely go astern, but the option is usually desirable for scale types.

Some car controllers employ resistances to drop current supply and hence speed, and switched circuits can incorporate car headlamp bulbs, lengths of electrical element etc. as resistances. Although such devices are simple, they have the disadvantage of wasting current; in effect, the boat is consuming the same current as it does at top speed, irrespective of its actual speed. In big models with a considerable power supply this may not be too important, but on average models where perhaps only 20 mins. at top speed is available between charges, a more efficient system is desirable. In general, it is far better to forget resistance-type speed control for boats.

Sail winches

Except for very tiny models, conventional servos are not suitable for controlling the sails of yachts and sailing craft. Exceptions are aerodynamically balanced sails, where virtually equal areas of a sail are exposed to the wind on either side of its pivot line. For normal fore and aft sails a winch mechanism is needed, and this is either of the drum or lever type. A drum winch is driven by an electric motor geared down to rotate four or five turns in two to three seconds, the drum diameter being such that 9-12in. (230-300mm) of sheeting line can be wound in or out. A lever unit is further geared down to swing an arm through about ¼ turn.

2·3 A commercial sail winch for larger yachts, employing a battery supply separate from the radio batteries.

Normally both sails of a standard Bermuda rig are operated simultaneously by one winch.

Commercial winches, plugging into the receiver, range in price from a little more than the cost of an average servo to three times that; they are proportional in operation and include limit switches at the extremes of movement. It is possible to construct such a winch, and articles have appeared in modelling journals but many home builders settle for a progressive winch, using the two microswitch reversible circuit shown earlier (fig 2.5) to control it, and a commercial motor with a Richard, MiniRichard or Pile gearbox fitted. Limit switches – simple contacts broken by knots in the sheeting line – are a desirable addition to the circuitry. This is a simple and inexpensive form of winch which is quite adequate for fun sailing.

It is still possible to see examples of winches operated by a push-rod from a servo, this usually moving a printed circuit board within the winch, causing current to flow through wipers and the winch to rotate until the wipers once more rest on insulated areas of the printed circuit board. To achieve four revolutions of the drum on, say, half a revolution of the wiper disc, the disc drive must be taken off earlier in the gear train at a point giving a ratio of 8:1, e.g. final drum gearing 20:1; disc gearing 160:1. One advantage of this system is that if a proportional servo is used, the response of the drum will also be proportional. Current practice, is however, to dispense with a control servo and plug a winch with a built-in amplifier directly into the receiver. Some of the smaller winches use the receiver power supply, but the larger ones usually require a separate battery.

2·4 *An example of a home-constructed winch using a Monoperm motor with attached Richard gearbox, for progressive sail control.*

Heavy duty servos

Apart from rudder control on large or high-powered models, there is little application for heavy-duty servos such as are used for undercarriage retraction, etc. on aircraft. Possibly one could be used for sail control on a small yacht, but since the cost often approximates that of a winch, there would be little point in buying one for this purpose when the winch would be more suitable. Applications in model boating are most likely to be for novelty effects where a conventional servo would not have adequate power but the movement required was insufficient to warrant the construction of a servo-switched motor unit. Lowering and raising the mast(s)/funnel on a river boat might be an example.

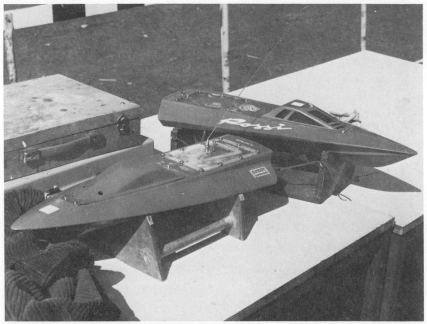

Two of Heinar Gundert's (W.G.) beautifully prepared models for F1-V2.5 and F1-V5. Note access hatches secured by blind nuts, also whip aerials.

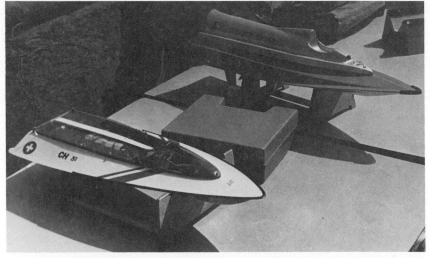

Two Swiss models, foreground an F1-E500 electric speed model with visible interior through clear hatch, background an F1-V15 i.c. speed model.

CHAPTER THREE

The Boats and Competition Classes

Radio is applied to almost every type of boat from converted plastic kits upward. To use conventional radio equipment as bought there is, of course, a minimum size, both for the physical accommodation of the radio and for the amount of weight involved, and this might be complicated by the basic power unit, the bulk and weight of which varies considerably between, say, a glow motor, electric power, and steam. Generalisations are difficult and must err in favour of the novice, but leaving steam aside as a special case, a rule of thumb is that if a model's length is multiplied by its beam and the result exceeds 120 (inch measurement) or 77,000 (millimetres), radio is likely to be a workable proposition.

6.5 x 18.5 in. = 120.25
165 x 470mm = 77,550

3.5 x 34 in. = 119
89 x 864mm = 76,896

FIG 3.1

Weight is more difficult, since while it would be simple to build, say a 20 × 6in. launch with a 1cc. engine and rudder and throttle control for under 2lb. (900g), the same boat with electric power and rudder and speed control is likely to be nearer 3lb. and will have a much inferior performance, especially if fitted with a low-price motor and dry batteries, which is what many beginners would do. Since radio weight remains standard, it is obviously better to build slightly larger electric models to gain initial experience, allowing a high proportion of total weight to be used for propulsion. Larger models make radio installation and access very much easier, too.

In discussing radio model boats, it is perhaps best to categorise them by the type of power used, and an immediate division occurs between mechanical propulsion and sail. Of these, the former can be broken down to the type of motor or engine used, so that fairly definite classes can be established.

Power boat types

There are many ways of dividing power boats into classes, most of them cross-referencing, but a useful starting point is whether the hull is of a planing or displacement type, i.e. whether it is intended to produce dynamic lift so that in motion it rises in the water and planes on or near the surface, or moves through the water with its waterline remaining effectively constant.

Planing and displacement hulls

Planing hulls, sometimes referred to as launch-type models, are normally very beamy and shallow, with a wide transom cutting off the hull squarely. They are

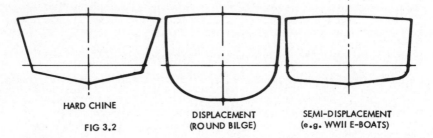

HARD CHINE

FIG 3.2

DISPLACEMENT
(ROUND BILGE)

SEMI-DISPLACEMENT
(e.g. WWII E-BOATS)

nearly always of hard chine form, that is, the sides and bottom panels join to form a pronounced corner along the chines. This corner helps to break away water flowing outward along the hull bottom which would otherwise tend to cling to the hull and greatly increase resistance.

A displacement hull, on the other hand, will normally be of round-bilge form, that is, a smooth curve from gunwale to keel with no corners, the sort of shape one associates with a ship, though it is found on many small craft, down to 8ft. dinghies. Such a hull has a critical speed above which the bow will rise and instability become apparent; there is a compromise form, termed a semi-displacement hull, with hard bilges (a small radius between sides and bottom rather than a hard chine) which produces a semi-planing performance and is occasionally found in launches and small craft up to about 120ft. length.

Unless based on a full-size planing hull, all scale models are likely to be of the displacement type – tugs, trawlers, liners, warships and so on. The exceptions are small, fast vessels such as service launches, offshore racers and high speed cabin cruisers which can give exciting performance in model sizes, though it might be mentioned that because of scaling effects on hydrodynamics, a true scale model will never be as fast as a functional model designed for speed. In general, electric power is normal for displacement hulls and internal combustion engines are used in planing models

Competition classes

All competition categories of radio power models except scale use planing hulls; this is another generalisation, but exceptions are rare at any level above club events. Radio competitions cover speed, steering and multi-boat racing for both i.c. and electric power, and scale steering and speciality events, and are most often run under Naviga rules. Naviga is the international body to which most model power boat nations belong and there are currently 33 classes of models for competitions, including tethered and free-running categories. For radio, the classes are:

Fl-Elkg.   – electric-powered speed models of up to 1kg. total weight
Fl-E500   – electric-powered speed models limited to 42v power supply
Fl-V2.5   – i.c.-engined speed models, 0-2.5c.c.
Fl-V5   – i.c.-engined speed models, 2.51-5c.c.
Fl-V15   – i.c.-engined speed models, 5.01-15c.c.
F2A   – scale models, length 70-110cm.
F2B   – scale models, length 110-170cm.
F2C   – scale models, length 170-250cm. (or $\frac{1}{100}$ scale if larger)
F3E   – functional steering models, electric powered
F3V   – functional steering models, i.c. powered
F6   – group or team demonstrations, 10 min. allowed
F7   – individual (scale) demonstrations, 10 min. allowed
FSR-3.5   – i.c. multi-racers, 0-3.5c.c.
FSR-6.5   – i.c. multi-racers, 3.51-6.54c.c.
FSR-15   – i.c. multi-racers, 6.55-15c.c.
FSR-35   – i.c. multi-racers spark ignition up to 35c.c.
FSR-E2kg   – electric multi-racers up to 2kg, 8 nicads, 15 min. races
FSR-E500   – electric multi-racers up to 42v max., 10 min. races.

3.1 Tugs are always popular for scale models. This excellent example is 37in (94cm) long.

3.2 Electric speed models have grown in popularity recently and can be bought in kit form, as this German example.

3·3 Warships attract the modeller who likes detail work; World War 2 frigates, corvettes and destroyers have particular appeal.

All courses except FSR are based on a triangle with 30m. sides; for steering events three buoys are added outside the basic triangle. Scale models (F2 classes) are judged statically and their scores added to the points gained from performance round the 'cloverleaf' course, which involves travelling astern through the last pair of buoys and a docking manoeuvre, the dock width and length being adjusted for the size of each competitor.

F3 boats are timed and the time built into their score, thus normally avoiding tied results. F6 and F7 are usually spectacular events, with all sorts of special effects and a high degree of ingenuity. FSR races are held around an M-shaped course, six or more boats at a time, the aim being the highest number of laps in 30 min. and the top six boats competing in a final.

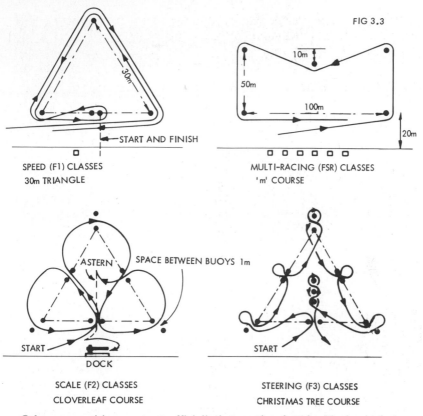

FIG 3.3

SPEED (F1) CLASSES
30m TRIANGLE

MULTI-RACING (FSR) CLASSES
'm' COURSE

SCALE (F2) CLASSES
CLOVERLEAF COURSE

STEERING (F3) CLASSES
CHRISTMAS TREE COURSE

Other competitions not yet officially international under Naviga include one and two hour multi-racing (or endurance racing), R/C three- or four-point hydroplanes, $\frac{1}{16}$ mile straight and 440 yard oval speed courses, and off-shore racing. In addition, there are many different types of event run by clubs, often as open competitions, particularly for scale models but also including relay racing, speed steering, towing, etc. for functional models.

For many years the most common single type of model was the semi-scale launch/cabin cruiser type with an i.c. engine, but during the 1970's electric scale models increased rapidly in popularity and now predominate. The expression 'semi-scale' is a convenient and generally accepted way of indicating that a model is not a scale subject but has something of the appearance of a possible full-size prototype rather than being purely functional.

Categories of scale models for competition may include 'scale' and 'stand-off scale', the former being a true-scale model accurate in every detail and the latter a model of scale type which looks correct at 7-10 yards distance but does not include every minute detail. Another useful definition is that a *boat* is a vessel that could be hoisted aboard a *ship*; thus an event could be specified as for scale ships, with scale boats as a separate competition.

33

Sail

Scale sailing models tend to a minority interest, due to some extent to the cube law ($\frac{1}{4}$ the length = $\frac{1}{16}$ the sail area = $\frac{1}{64}$ the displacement, or pro rata) which makes stability sometimes marginal, but also because scale sailing vessels can have a disappointing performance. Fore and aft rigs are reasonable, though almost always need a false keel for stability and efficient sailing, but square riggers will usually only travel off the wind. A square rigger can be trimmed to work to windward, but frequently the leeway it makes exceeds the distance moved to windward.

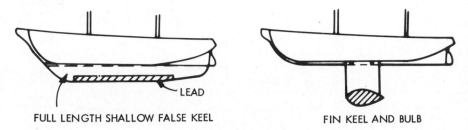

FULL LENGTH SHALLOW FALSE KEEL   FIN KEEL AND BULB

FIG 3.4

This is not to say that scale sailers are impossible, or do not give pleasure, but they can be a little tricky unless of a fairly basic fore and aft type. The larger the model the fewer the problems, and the better the performance, is worth remembering. Very few successful models are seen without additional side area below the keel, which also allows ballast to be carried lower; the alternatives are to distort the hull or drastically reduce the sail carried, and of the choices the false keel is the most helpful and least objectionable. Square sails are best left furled and a model sailed on fore and aft sails only.

3·4 A scale schooner at the Round Pond, Kensington. The fin and bulb extension keel is noticeable.

Competition classes

An indication of numbers involved in scale sailing models is given by the fact that there are normally only three major regattas for them in Britain during a season, compared with 150 or more for racing yachts and 250 or more for power boats.

Radio controlled yacht racing has been a major growth area in recent years all over the world. While there are several kit models of, usually, scale-type appearance, racing is confined to classes with strict rules accepted internationally, and an intending builder is advised to ensure that any model he fancies does in fact fit one of the recognised classes. Perhaps more than in any other aspect of modelling, sailing against other models is what makes model yachting enjoyable.

In order of popularity, the international classes are:

Marblehead (RM). Limited to an overall length of 50in. plus or minus ¼in. and 800sq. in. of sail, this class is sailed in all yachting countries. The yachts average 12-18lb. and there are no limits on beam or draught.

10-rater (RIOR). Between 6 and 7ft. long and very slim, the formula for these boats relates sail area and waterline length, the two main speed factors. A 55in. w.l. boat can have 1363sq. in. of sail, a 60in. one 1250, a 65in. one 1153sq. in. Displacement is usually 20-24lb. and the yachts are very fast. They are sailed in most countries.

East Coast 12 metre (EC12). This is a one-design class using a scale 12m hull 59in. in length, very popular in the U.S.A., where it originated, but not widely used elsewhere, though it was only adopted as an international class in 1980. Sail area is about 1200sq. in. and maximum displacement 26½lb., which, with a draught of 8¾in. maximum makes for a rather tender boat in any wind.

A class (RA). The largest class, the rating rule allows a wide variation in displacement, from 20-90lb. An average yacht would be perhaps 52lb. on a 55in. waterline (75in. overall), 12in. draught and 1500sq. in. of sail, but there is now a tendency towards lighter boats, accepting the reduced sail allowed for them. While fleets exist in several countries, the size has inhibited growth for radio purposes, portability being a problem except where clubs have storage facilities at the waterside.

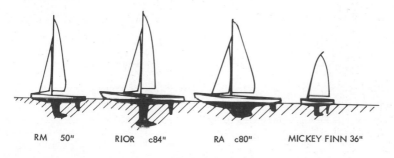

RM   50"   RIOR   c84"   RA   c80"   MICKEY FINN 36"

FIG 3.5

Two other classes, the 6m. and 10/40, are recognised, but currently there is no radio activity in the former and little in the latter. So far no multi-hull class is recognised in any country, though there have been experimental classes and individual models. In Britain there is the 36in. Restricted class ($36 \times 9 \times 11$in. max., 12lb. max.) and two one-designs, the 36in. *Mickey Finn* una rig and the 1.5m, plus the unofficial but popular tiny 575 kit model, $17\frac{1}{2}$in. long.

Radio yachts are raced in heats of 6-12 boats, with a 1-minute countdown to the start and, usually, two laps of a triangular course per heat. Sailing rules are virtually identical with full-size rules and points are scored in each heat and totalled at the end of the day. An alternative is to divide the entry into fleets and after each heat move the top two boats up into the next fleet while the bottom two move down; this gives closer racing and is popular with most competitors.

The yachts are Bermuda rigged, that is, with two triangular sails, the jib and the main. Contrary to full-size practice, both sails are fitted to booms and the booms are controlled by the radio, usually synchronously, i.e. both sails are controlled by a single winch and are so arranged as to retain their initial relationship throughout their travel. The same arrangement can be used with other fore and aft rigged models where booms are fitted, since it is only necessary to make the attachment point of the sheet on each boom an identical distance from the boom pivot point. With loose-footed sails, where the sheet is attached to the sail clew, it is necessary to use a stepped drum on the winch (or different attachment points on a lever winch) in order to ensure that all sails move through the same angle irrespective of the distance of the clew from the pivot point of each arc. Establishing the amount of sheet travel required for each sail, and hence drum diameter or lever radius, is a matter of simple geometry.

Newcomers to sail are often unaware that there is only one correct sail setting for each course to be sailed and that proper control over a sailing model therefore requires infinite adjustment of the sails. A boat should sail on its sail trim alone and the application of rudder slows it down. While it is possible to sail a model round a lake with rudder control only, the sails will not be working efficiently for most of the time and the amount of rudder needed will be excessive.

3·5 Radio Marbleheads and 10-raters at an international regatta. Japanese, German and French boats in the foreground.

3·6 Coming up to the start line in a radio Marblehead race in strong wind at Fleetwood.

3.7 A radio 10-rater being checked by Chris Dicks, one of England's top vane and radio skippers

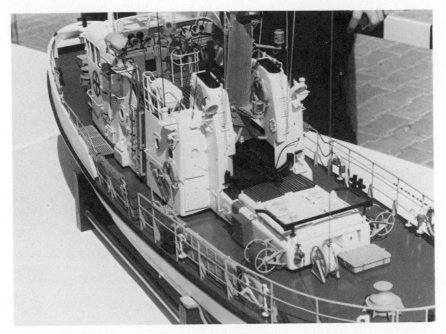

Access to the interior has to be kept in mind during construction. With both the German lifeboat above and the Japanese gunboat below it is clear that the main superstructure lifts off – a difficult joint to hide.

CHAPTER FOUR

The Construction

Model boating enthusiasts divide into those whose pleasure comes from running a model and/or competing and those who enjoy building more than, or as much as, operation. Thus despite modern materials which allow pre-formed hulls, etc., there is still a huge following for traditional wood construction. A majority of functional competition models, power and sail, use GRP (glass fibre) construction, usually from commercially available hulls, though it is not uncommon, particularly with yachtsmen, for individuals to mould their own hulls. Scale models may employ GRP, or vacuum-formed styrene hulls (now provided in many kits) but most are likely to be of wood; much the same applies to sport models, built for the fun of running a boat rather than competition participation.

Hard chine hulls

These are almost always of wood, except for competition speed models and multi-racers and, occasionally, styrene hulls in kits. The form lends itself to ply construction, so that apart from hydrodynamic suitability, such hulls are relatively quick, simple and inexpensive to build and availability of materials is seldom a problem. A basic frame is built with a sawn or pre-laminated keel member to which bulkheads are attached and chine stringers and inwales (at the deck/side junction) are added. Sometimes self-aligning features are built in; the 'proper' way is to mount the bulkheads on a jig so that no distortion can creep in. Laminated chines and inwales, to eliminate heavy bending stresses during construction, allow a careful builder to produce a true hull without using a jig, but where absolute symmetry is essential, as in a yacht, a jig is recommended.

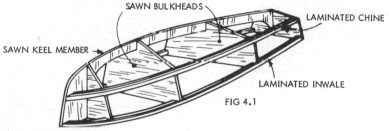

SAWN BULKHEADS

LAMINATED CHINE

SAWN KEEL MEMBER

LAMINATED INWALE

FIG 4.1

39

When the hull is 'in frame' the keel, chines and inwales are chiselled or planed to match the sections and the areas to be covered sanded with a long glasspaper block to ensure a fair seating for the skin panels, which are usually cut from thin ply. The normal chine hull will have two bottom and two side panels and a deck, and it is usual to cut thick paper or thin card templates to establish the shapes. In many cases the bottom panels will butt together over the keel along the centre line, but occasionally the keel carries doublers providing a lodging for the panels, leaving the main keel protruding. Doublers are required each side in the area of propeller and rudder tubes, or where the fin keel of a yacht is fitted, and it is usually best to make up the keel complete with propeller tube, etc. as the first unit, before securing bulkheads, etc. The skins are normally pinned and clamped in place (clothes pegs or bulldog clips are suitable) and the pins etc. removed once the glue has set.

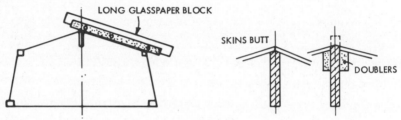

FIG 4.2

Frequently the most awkward bit is merging the skin-panels along the chine at the bow; the side-skins usually overlap the bottom ones along the chine, but where the angle between them becomes very shallow, the overlap should be stopped and a butt joint made, as illustrated. Sharp curves, such as on a blunt bow, are better planked with narrow vertical strips, or a carved block used, terminating the main skins on a suitably positioned bulkhead.

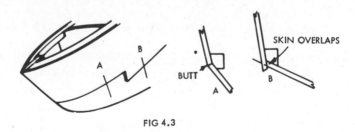

FIG 4.3

Bread and butter construction

Carved hulls from a single block are not normally seen nowadays, and are not recommended. Bread and butter construction, laminating shaped planks (the bread) with glue (the butter) is infinitely better and more economical, and it is really the simplest way of achieving some of the more complex shapes such as, perhaps, the stern tuck of a tug or the propeller tunnels of a lifeboat.

Drawings of a ship or yacht normally include a body plan, consisting of sections vertically through the hull at equally spaced intervals, superimposed on common axes (the centre line and the load waterline) with the forward half of the hull on one side of the centre line and the stern half on the other. If the available timber thickness is drawn across the body plan, the width of each plank at each section station can be taken off with dividers. The timber is thus marked with the station positions, the corresponding widths spotted, and a line curving through the spots gives the shape of the lamination to be sawn. It is more economic to mark half-planks, joined on the centre line.

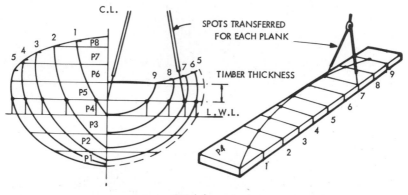

FIG 4.4

Work can be saved by cutting out the centres of all planks (except the bottom one) and this can be plotted by tracing each section and drawing on it the required hull thickness, which then provides the measurements for an inside cutting line, transferred to the timber in the same way. Note that the plank thicknesses must be drawn on the hull profile (sheer plan) at bow and stern, to establish the actual length of each plank and the point at which the inside cutting line must terminate.

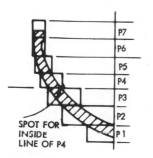

SPOT FOR
INSIDE
LINE OF P4

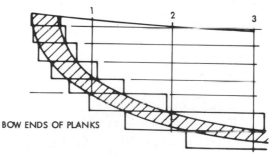

BOW ENDS OF PLANKS

FIG 4.5

41

Mark the station lines clearly on all planks, since these are used to align the planks when they are glued up into a block. It is best to drill two holes through the bottom plank to bolt the block firmly to the bench while carving, plugging them on completion.

It is appreciated that this process can seem puzzling at first to a beginner, but a little thought will clarify it. Even if a full bread and butter hull is not thought likely, the method remains the same for laminated bows and sterns with planked midships sections, which is one of the commonest ways of building a scale hull.

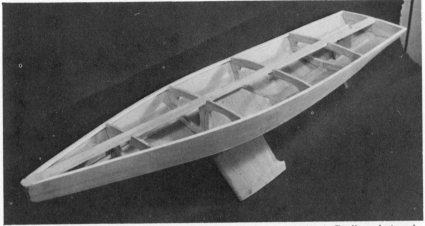

4·1 A simple 36in hard chine yacht hull of balsa and ply construction is Gosling, *designed by the author.*

4·2 Bread and butter construction at the stern with planking midships is easiest and most economical in many cases.

42

Planked hulls

These can be plank-on-frame or rib and plank. The former is rather like the hard chine method described above, without chine stringers, and with the bulk-heads (or frames) remaining part of the completed hull. With rib and plank, undersize bulkheads, called shadows or moulds, are mounted on a jig and ribs – narrow strips of thin ply – pinned round them. After planking, the shadows are unscrewed from the jig, the hull lifted off and the shadows twisted out, leaving a shell of planking with just the ribs internally. Deck beams are cut and fitted across between the inwales to preserve the correct shape and support the deck. It is, of course, possible to remove the bulkheads in a hard chine hull to achieve a similar result.

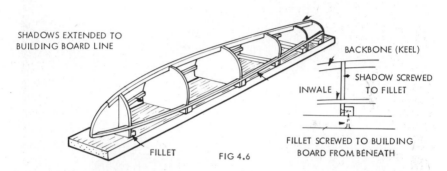

SHADOWS EXTENDED TO
BUILDING BOARD LINE

BACKBONE (KEEL)

INWALE

SHADOW SCREWED
TO FILLET

FILLET

FIG 4.6

FILLET SCREWED TO BUILDING
BOARD FROM BENEATH

Plank-on-frame is usually used for power boats, with bulkheads left in, although frequently the centres of the bulkheads are fretted out; a solid member is a bulkhead, but if the centre is removed it becomes a frame. Rib and plank construction is most frequently used for yacht hulls. The shapes for the bulkheads/frames are obtained by tracing the sections from the body plan and reducing them by the finished thickness of the planking; in the case of shadows, they are reduced by the planking thickness plus the rib thickness.

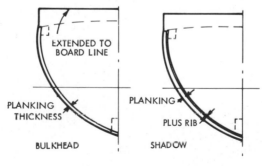

EXTENDED TO
BOARD LINE

PLANKING
THICKNESS

BULKHEAD

PLANKING

PLUS RIB

SHADOW

FIG 4.7

Individual planks can be plotted by dividing the largest hull section into the number of planks required, then a chosen section towards each end by the same number. Straight lines drawn over the body plan connecting corresponding points on large and chosen sections will then give plank widths at each intermediate section, and if the lines are extended to the centre line they will indicate which planks will terminate on the keel or backbone rather than run right to the extreme ends of the hull. The chosen sections towards the ends should fall immediately before any major alteration of the hull shape.

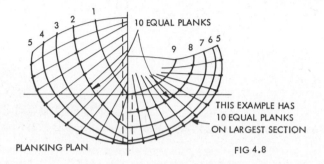

PLANKING PLAN

FIG 4.8

It is not possible to go too deeply into the details of planking in this book, but it is not difficult to find other reference books covering the subject more fully.

GRP hulls

These are produced in a mould which is in turn made on a 'plug', which is a solid carved model. To make the plug, bread and butter construction is usually employed, but there is no need for internal hollowing. Since it is used only to make the mould, reclaimed timber is often used; the vital thing is to have an accurate shape with a flawless finish, as any imperfections will be incorporated into the mould and thence into the finished moulding. A usual practice is to carve the shape (often extended an inch or so above deck level) and apply several coats of polyester resin, perhaps with a filler powder added. This can then be rubbed down and polished until a perfect surface is obtained.

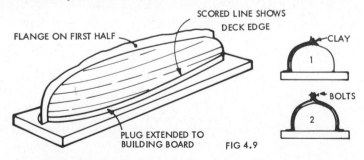

FIG 4.9

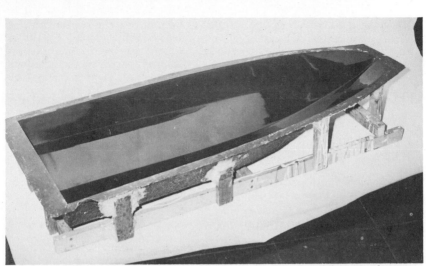

4·3 *A one-piece mould for a glass-fibre hull. Note the high finish and the external reinforcement*

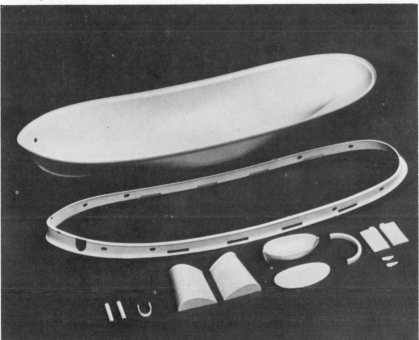

4·4 *Vacuum-formed parts for a Graupner* Bugsier *tug kit include small parts as well as the main hull etc.*

Most moulds are made in two halves by building a wall of clay or Plasticine along the centre line of the plug, spraying the first half with release agent (an alternative is several thorough coats of wax polish, well rubbed) and then painting on a gel coat, a special surfacing resin. When set and only slightly tacky, a layer of glass mat is applied, stippled through with polyester resin and avoiding air bubbles or dry (opaque) patches. A second layer is then added, possibly a third for a large mould; strips of wood or wire etc. can be bonded on to stiffen the mould.

The clay wall is then removed and all traces cleaned away. Release agent is applied to the second half and the flange of the half-mould, and the process repeated. Holes are drilled along the centre flanges before removing the half-moulds from the plug. This may take a little persuasion and possibly the insertion of thin strips of ply where possible between mould face and plug, taking care not to introduce scratches.

Check the mould for imperfections, bolt the halves together along the flange, polish well, apply release agent, gel coat, glass mat or cloth etc. as before. Keen builders weigh resin and cloth to the intended weight of the hull, then spread these amounts of material only. The glass must be thoroughly wetted out, but excess resin simply adds weight and no strength. Remove moulding from mould as before.

Always mix resins exactly to the makers' instructions and work in the recommended temperature. Green (new) mouldings may appear set but can distort; curing takes a day or two and the mould should be left on the plug or the moulding in the mould to avoid distortion. GRP is very much easier to cut while green, but replace in the mould afterwards. Once cured, accessories can best be bonded in place by thoroughly roughening the surface and using epoxy resin or, better, reinforced by draping with patches of scrap glass mat and wetting in thoroughly with resin. The moulding surface must, however, be scored or roughened down to the glass fibres to ensure a really firm bond.

Colour pigments can be added to the resin while moulding, but if a painted finish is required, wash thoroughly with detergent, and rinse, then cloud the surface with a worn piece of fine carborundum paper. Use polyurethane or epoxide paints.

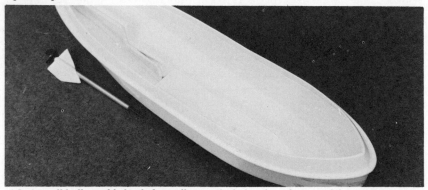

4·5 A small hull moulded in halves, allowing the bulwarks to be part of the main mouldings.

Vacuum-formed hulls

These are normally outside the province of the individual, since they need a vacuum-forming machine, but polystyrene card can be used to make complete hulls, radio boxes, or many fittings. For hulls it is best to think of one-plane curves, like plywood, but for fittings it can be heat-moulded. An undersize male mould and an exact size shape cut in a piece of ply are needed; the plastic is pinned over the cut-out and warmed in front of an electric fire until it suddenly looks floppy, when the male mould can be pressed against it. Further warmings may be necessary to complete a deepish draw.

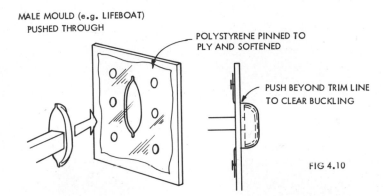

MALE MOULD (e.g. LIFEBOAT) PUSHED THROUGH

POLYSTYRENE PINNED TO PLY AND SOFTENED

PUSH BEYOND TRIM LINE TO CLEAR BUCKLING

FIG 4.10

Polystyrene adhesives, either in tubes or bottled liquid (methylethyl ketane) rely on a solvent action and must thus be used sparingly. When the first application is dry, a second coat can be applied. Liquid cement is safer, since it can be applied with a small brush and capillary action will draw it along a joint line, but a second coat brushed along the joint is usually desirable. Sawing or filing the material seems to generate a good deal of static electricity and the best way to cut it is to make a continuous and reasonably pronounced score line with a sharp blade, when the plastic will snap cleanly through on gently flexing.

Sources

Most modellers start with a kit, usually of hard chine wood construction or possibly with a GRP or styrene hull. Advantages of kits are that they should contain all necessary parts, with the usual exceptions of short-life adhesives and paints, and that in most cases much of the hard work and trickier bits have been done by the maker. Few power kits contain the motor, and unless the recommended motor, radio, etc. are used, it may be necessary to make alterations. The chance of selecting material is also largely sacrificed, though modern kits from reputable manufacturers rarely include indifferent materials. The costs of prefabrication, die-cutting, boxing, advertising, etc. are reflected in the price of the kit, but this is obviously no deterrent to most modellers, considering the vast numbers of kits sold.

Some saving can be made by buying a hull and plans, and a selection of GRP (and sometimes styrene) hulls is readily available for competition, scale and sport boats and, to a lesser extent, yachts to racing classes, although it is sometimes necessary to buy by mail order, since no model shop is likely to stock more than a few examples. The necessary materials for completion can then be bought locally, unless special fittings are called for, when these too may have to be bought postally. Membership of a club is a great help in this sort of project.

The most economic form of building is to work to a plan, which again will probably mean sending away for one. Because the capital outlay to produce plans is far less than is needed for kits and even hulls, a very wide choice is available. Most make use of standard materials and give an indication of the degree of experience necessary in catalogues or descriptive literature. The builder is faced with the necessity of tracing parts out and sawing or cutting, or, if the plan is for experts and shows no construction detail, plotting out the various components, but this takes a much shorter time than is often imagined and, once the pieces are cut, assembly and finishing takes no longer than a wood kit.

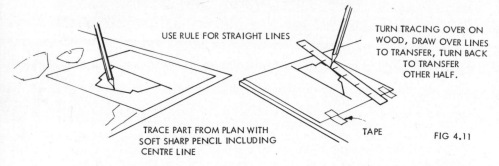

USE RULE FOR STRAIGHT LINES

TURN TRACING OVER ON WOOD, DRAW OVER LINES TO TRANSFER, TURN BACK TO TRANSFER OTHER HALF.

TRACE PART FROM PLAN WITH SOFT SHARP PENCIL INCLUDING CENTRE LINE

TAPE

FIG 4.11

Original designs are frequently seen, but the successful ones are almost invariably produced by modellers who have gained experience with kits and plans, both in construction and operation, and therefore have knowledge of the things to avoid. A beginner is advised to use a proved design to familiarise himself with the basics; some of us are still learning after fifty years of practical modelling!

CHAPTER FIVE

The Power

The various types of power can be classified under the following headings: Internal combustion engines, electric motors, steam and sail. The advantage of internal combustion engines is that, provided adequate fuel is on hand, unlimited running is possible. The main disadvantage is noise, which in these pollution-conscious times is a serious matter. For competition purposes, noise emission limits are imposed, but even for sport running, common-sense dictates that to avoid causing a nuisance and to ensure continued availability of facilities, every effort should be made to reduce noise to an acceptable level.

Internal combustion engines

There are three now traditional types of i.c. engine, diesel, glow and spark-ignition, but with an increasing number of four-stroke glow engines we may eventually see a fourth category.

Diesels, or, more accurately, compression-ignition engines, were developed during the 1939-45 war and enjoyed a considerable vogue in Europe for some years. Nowadays they are found only in smaller sizes, 2½-3½c.c. being most popular for boats, although marine versions are available from .5 to 5c.c. and they thus cover launch-types of around 18-40in. length. They have the advantage of requiring only fuel and a starting cord, but the disadvantage of a relatively limited throttle response; few below 1½c.c. are ever fitted with throttles.

They normally run a little cooler than glow-engines and use a fuel composed of ether, paraffin (kerosene) and castor oil, or a similar mixture, sometimes with mineral oil. With adjustable compression and fuel control needle, establishing starting/running settings can be puzzling to newcomers, but if the engine is flooded and the compression backed off a turn or more, pulling it over and increasing the compression a quarter turn at a time should soon produce firing or short bursts of running. It will then be possible to establish whether the engine is stopping because it is getting too much fuel (close needle gradually) or too little (open needle a quarter turn at a time). Always turn the flywheel by hand before using a cord or starter to check that it is free to turn; too much fuel/compression can lock the engine. Throttles for diesels are usually simple barrel types, though a screw may be fitted to prevent full closure and can be adjusted to obtain a steady

5·1 *Diesel and glowplug engines of similar (3½cc) capacity, the diesel with rear disc induction and the glowplug with front rotary.*

slow-running speed. It will probably be found best to have the throttle at least three-quarters open for initial starts.

Two-stroke glowplug engines are by far the most popular form of i.c. power, developed in the U.S.A. during World War II and usually lighter, faster 'revving' and more responsive to throttle alterations than diesels. There are four popular sizes, 2.5, 3.5, 6.54 and 10c.c. (.15, .19, .40 and .60cu. in.) but marine glow-engines can be found in most sizes from 1 to 15c.c. Like diesels, they are usually used in launch-type models (q.v.) and only rarely in displacement hulls. Disadvantages are the chance of burning out glowplugs, the need to carry an accumulator for starting purposes, the exhaust temperature (which is considerably higher than a diesel) and the need for protection of the boat structure and finish against fuel attack. Boat size can be 18-54in., depending on the engine size and purpose of the boat, i.e. scale or scale-type launch, speed model, multi-racer, etc; the vast majority of competition speed and racing models use glow-engines. The new trend to four-stroke glow-engines tends to overlap the spark-ignition category rather than two-stroke glow classes.

Glowplug engines have only a needle control and if flooded should produce an audible sizzling sound. Close the needle and spin over to expel excess fuel, if necessary removing the glowplug. The glowplug filament should be seen to glow brightly when the starting battery is connected; failure to start can only be due to lack of fuel or a plug which fails to glow adequately (low battery) or is burned out (replace plug). Check whether the plug is for 1.5 or 2v. operation and that it is of a type recommended by the engine manufacturer.

Glow fuel is a mixture of methanol (methyl alcohol) and oil, often castor but sometimes mineral. Nitro-methane is sometimes added, and some engines, especially small ones or some foreign makes, will not run satisfactorily without 5 or

50

5·2 *Italian racing engines are among the best. This is a 10cc Rossi with rear disc induction and glowplug ignition.*

10% nitro in the fuel, whatever type of plug is fitted, so use the recommended fuel or seek advice from your model shop or an experienced modeller.

Throttle response is better on a glow-engine than a diesel, but a more sophisticated carburettor may be fitted with different jets for idling and fast running. Adjustment is more a matter of patience than skill; start and adjust the setting at more than half throttle initially and tackle idling speed after satisfactory top end performance is achieved. All motors benefit from a running-in period, running for half- to one-minute periods on a rich mixture at first, then operating on an oil-rich mixture at less than top speed for half an hour or so on the water.

Spark-ignition is by far the oldest form of internal combustion engine, going back to the early 1900's; present day engines tend to be 15c.c. or upwards, to the international model limit of 35c.c., and may be two or four-stroke. There are a few smaller examples in use but most seen today are adaptations of industrial motors (chainsaws, generating sets, etc.) and the result is that they require biggish boats, typically launch-types over 4ft., speed models of 3ft. 6in. or so, or scale tugs and trawlers perhaps 5ft. or more in length. Fuel is normally petrol (gasoline) and the old coil and condenser ignition system has gradually given way to magnetos, usually of the flywheel type.

Provided fuel is getting through and a good spark is occurring, running should offer few problems. Water may affect the ignition if for some reason the engine gets wet, and there may be a faint chance of radio interference from the spark. A screened lead and suppressor as used for car plugs will normally be adequate with modern radio equipment. Like glow-engines, the exhaust gases are hot, and brazed joints or silicon tube connections are needed in the exhaust system, at least close to the engine.

Electric power

More boats are fitted with electric power than all other forms of power combined, ranging from the smallest, perhaps a converted plastic kit model with a tiny motor and a single dry cell, to vast models 12-13ft. long, sometimes carrying two or three automobile batteries. Along the way are sophisticated speed models packed with nickel cadmium cells, but the majority of electric models are scale, or at least of scale type, using permanent magnet motors and increasingly, carrying rechargeable cells. Advantages are the range of sizes, clean, quiet running, instant starting and ease of reversing for astern manoeuvring. Disadvantages may be low power/weight ratio (i.e. motor and batteries weigh a lot for the power produced) and running time limited to the battery capacity; the larger the model, the less these considerations apply. In initial expense, there is little difference between the average i.c. engine and the average electric installation with rechargeable cells, but running cost are much lower with electrics.

5·3 Nicad packs, 8 cell and 16 cell (9·6 & 19·2v, 1·2a/h) with accessories for aircraft or boat use.

Electric motors are available in wide variety, and nowadays all but a very few are of the permanent magnet type, which means that they are reversible by simply reversing polarity. The power of an electric motor is usually expressed in watts, that is, the input voltage × the current drawn in amps, but electrical efficiency varies considerably. As a generalisation, efficient model motors cost more but will do more work on the same battery supply. This means increased performance from higher speed, longer running, or lighter weight, since a smaller battery could be carried to achieve the same power output as a cheaper motor.

Conventional dry cells have been losing favour in recent years, since they have become relatively expensive and compare poorly with nickel cadmium (nicad) cells in the amount of power they can supply against their weight. Exceptions can be found where weight is not an important factor; for example, a large scale model which would need ballast anyway could well use lantern batteries and achieve

52

many hours of running. However, even in such cases the tendency is to use rechargeable batteries, perhaps lead acid or nickel iron or, more probably nowadays, nicads. Nickel iron cells are heavy but stand enormous abuse, lead acid not quite so weighty but requiring regular attention to avoid deterioration.

Battery capacity is expressed in ampere/hours, indicating the recommended maximum discharge rate for a one-hour period. Thus one of the commonest lead acid batteries at 6v., 4a/h will theoretically deliver four amps for one hour at six volts, or one amp for four hours, etc. In fact it would be characteristic for such a battery to require rests to achieve approaching one hour's supply at four amps, i.e. it could not be continually used for one hour at such a discharge rate. On the other hand, with intermittent use it would probably supply one amp for perhaps five hours or more, rather than four.

The a/h capacity is also a guide to charge rates, the maximum charge recommended being $\frac{1}{10}$ of the a/h figure for, in theory, ten hours, (the 10-hour rate) though in practice an hour or two more would be needed. The 20-hour rate means $\frac{1}{20}$ of the capacity for 20 hours.

Nicads are now almost universal, since they are lighter for the same capacity and are totally sealed, requiring no maintenance other than cleanliness. They also stand a lot of abuse, since they can be discharged at a higher rate than theory dictates and can be neglected for months; two or three charge/discharge cycles will restore them. Standard practice is to charge for fourteen hours at the ten hour rate from flat, which is effectively when voltage drops to 1.1v per cell. In fact the charged voltage is only 1.2v per cell, against 1.5v for dry-cells or lead acids, but the discharge curve is flat, i.e. where a dry-cell delivers a gradually decreasing current, a nicad supplies full current until almost completely discharged, when the current drops quite rapidly.

Another type of nicad is the fast charge vented cell. These can be discharged at something like ten times the nominal rate and charged similarly, but it is essential to be precise in timing when charging, since pumping in a heavy charge for even a minute too long can cause severe damage. Such cells are used in fast electric competition models and are still something of a specialist area; most users are equipped with meters and special chargers which discharge cell packs before recharging so that safe charging in a matter of 10-12 minutes can be regularly achieved.

Buying nicads is initially more expensive, naturally; they cost several times the price of dry-cells and a charger is also required. However, they will last for, literally, years – at least 1000 cycles with only moderate care – and the total cost is likely to be less than the average i.c. motor.

Steam

Since the mid-1970s steam power has shown something of a revival, although it is still a comparatively specialist interest. Steam plant capable of good power output is fairly expensive to buy, though cheap to run, but it is also simpler to make, given some ability with tools, than other forms of mechanical power. Thus a good many units built in the home workshop are seen at regattas, in company with several smaller, less powerful plants which have been introduced ready-built

53

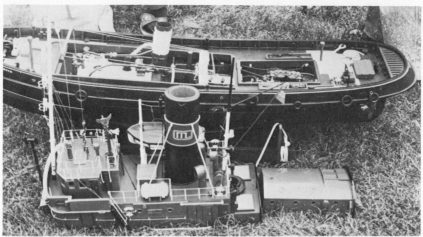

5·4 Probably more steam engines are fitted in tugs than any other models. This one is a 54in. Cervia.

within the last few years. In general, steam units are relatively heavy for the power produced and, in their simplest forms, are not easily controlled by radio. There is therefore a tendency to divide steam models in a simple category, with no radio or rudder control only, and with simple boilers and burners, and the more complex, larger models with multi-cylinder engines, pressurised blowlamps, reversing gear, feed pumps and other engineering-type ancillaries. Boats with the simpler units

5·5 A simple oscillating engine, the Unit 1, with boiler etc, suitable for up to 30in models.

may weigh only 4-5lb., the more complex 30-40lb. or more. Any interference with the steam supply on a simple single-cylinder engine is likely to cause the engine to stop in such a position that a manual turn on the flywheel is needed to restart it, so that radio control is only normally applied to engines with two or more cylinders and a self-starting capability.

Steam engines and boilers tend to be a source of heat and condensate may splatter around at times. It follows that radio should be kept at a distance and well protected, not really difficult measures. Actual engine control, of speed and direction, will depend on the valve gear, etc. of the engine and its linkage, and as already mentioned, a multi-cylinder engine unable to stop dead is essential. Many engines have speed and direction of rotation controlled by a reversing lever and a regulator lever, both of which can be operated simply enough by radio. Basically any control which can be adjusted manually can be connected to a servo, provided the force required is within the servo's power.

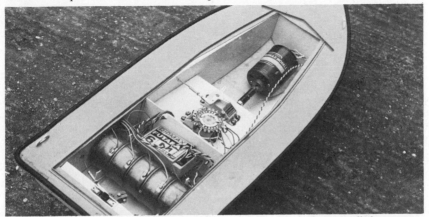

5·6 *A fast electric boat fitted with large cylindrical nicads and a radio-controlled rotary switching gear*

Propeller selection

The best propeller for a particular engine/hull combination depends on so many variables that trials are the only way to find out, but for average models the following suggestions provide practical starting points. For best results, the next smallest and largest diameters and/or the next lowest or greatest pitch should be tried, though unless the last ounce of performance is sought, a choice of three propellers will satisfy most modellers. Pitch is the theoretical distance moved forward by the propeller in one revolution, but in practice there is always a certain amount of 'slip'.

Diesel engines generally deliver maximum power at lower r.p.m. than glow-engines and thus usually have a slightly larger propeller. Electric motors are tabled under wattage (watts = amps × volts). Scale models may require three- or four-blade propellers but otherwise two-bladers are rcommended. Diameters are given in inches and millimetres, relating to sizes commercially available, and plain or standard pitch is intended.

ELECTRIC

Watts	2-blade		3-blade		Approx. shaft size
	in.	mm	in.	mm	
5	7/8	—	3/4	—	2mm-3/32in.
10	1-1¼	30	1-1⅛	30	3/32in.-4BA
15	1¼-1½	30-40	1¼-1⅜	30-35	4BA
20	1½-1¾	40-45	1½	40	4BA
30	1¾-2	45-50	1¾	45	4BA
50	2-2½	50-65	1¾-2¼	45-60	4BA-2BA

DIESEL

c.c.	cu. in.	Average sport model		Fast model		Approx. shaft size.
		in.	mm	in.	mm	
0.5	0.03	1	30	1	25-30	4BA
0.8	0.049	1-1¼	30	1-1¼	30	4BA
1.0	0.06	1¼-1⅜	30-35	1¼	30-35	4BA
1.5	0.09	1⅜-1½	35-40	1¼-1⅜	30-35	4BA
2.5	0.15	1½-1¾	40-45	1⅜-1½	35-40	4BA
3.5	0.21	1½-2	40-50	1⅜-1¾	35-45	4BA-2BA
5.0	0.29	1¾-2¼	45-55	1¾-2	45-50	2BA

GLOW

c.c.	cu. in.	in.	mm	in.	mm	shaft size
0.8	0.049	¾-7/8	—	¾-7/8	—	4BA
1.0	0.06	1-1¼	30	7/8-1¼	30	4BA
1.5	0.09	1⅛-1⅜	30-35	1⅛-1⅜	30-35	4BA
2.5	0.15	1¼-1½	35	1¼-1⅜	30-35	4BA
3.5	0.21	1⅜-1½	35-40	1⅜-1½	35-40	2BA
5.0	0.29	1½-2	40-50	1½-1¾	40-45	2BA
6.5	0.40	2-2¼	50-55	1¾-2	45-50	2BA
10.0	0.61	2¼-2½	55-65	2-2½	50-65	2BA

Notes.
1. Figures can only be a guide, since power or running speed of engines vary considerably and weight and type of hull, resistance etc. affect propeller performance.
2. Plain or standard pitch is approx. 1.2-1.3 × diameter, average high or X pitch 1.4-1.6 × diameter.
3. 4BA is .142in., approx. 3.5mm, 2BA is .185in., approx. 4.5mm, referring to diameter of plain (i.e. unthreaded) shaft.

Noise suppression

Most of the noise problem comes from the high-frequency sound waves generated by the engine, so that the higher the engine revolutions the greater is the problem. Besides the exhaust emission, the whole engine radiates sound waves; indeed, a metal mounting bonded to a GRP hull ensures that the whole boat resonates. Briefly, the steps which should be taken are: (1) lead the exhaust from the engine into an expansion chamber, preferably changing direction of flow; the exit pipe should be about the same sectional area as the exhaust ports. A restrictor allowing the outlet area to be reduced is effective, but may reduce engine r.p.m. (2) Keep the exhaust system inside the hull, preferably passing

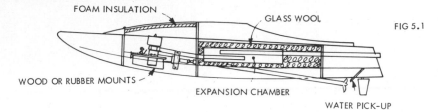

FOAM INSULATION GLASS WOOL FIG 5.1

WOOD OR RUBBER MOUNTS

EXPANSION CHAMBER

WATER PICK-UP

most of its length through a box packed with glass wool. (3) Fit a hatch over the engine bay, possibly lined underneath with foam rubber to absorb engine radiation. (4) Incorporate some wood (a ply mount?) in the engine installation so that a lot of vibration is dampened and not transferred to the hull. The ultimate is hard rubber mounting blocks (available commercially) though these also require a flexible propeller shaft or propeller tube mounting (also available).

Tuned pipe exhausts work just as effectively when boxed-in and led into a silencer chamber; 'quiet pipes' can be bought. They boost power by matching exhaust pressure waves to the position and speed of the piston, sucking exhaust and eventually fresh mixture from the exhaust port, then pushing part of it back as the transfer port closes, thus increasing the volume of fresh mixture in the cylinder as the exhaust port closes. Length is fairly critical, since the range of revolutions over which they will work is limited; normally a silicon tube connector is used between engine exhaust stub and tuned pipe to allow length adjustment.

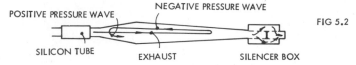

POSITIVE PRESSURE WAVE NEGATIVE PRESSURE WAVE FIG 5.2

SILICON TUBE EXHAUST SILENCER BOX

Fuel tanks

Tanks for fuel are usually fairly simple, the main problem being to mount them sufficiently low to ensure that the fuel level when full is no higher than the engines's spray bar or carburettor fuel inlet. This is not normally difficult with a small tank, but for extended running, large-capacity tanks are needed and these have to be fairly shallow and flat, which means that the fuel can surge about in the tank as the boat turns. To ensure a steady fuel supply, a small sump is usually fitted beneath the tank, with a form of baffle between it and the main tank, and the fuel feed is taken from this. A simple method is to drill a series of 3 or 4mm holes close together in one side of a metal fuel can and to solder over the holes the screw cap from another can. A brass tube soldered into this cap provides the feed pipe. Filler and vent pipes are then soldered in the top side.

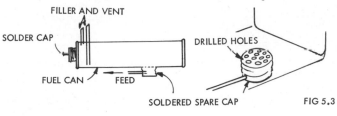

FILLER AND VENT

SOLDER CAP DRILLED HOLES

FUEL CAN FEED

SOLDERED SPARE CAP FIG 5.3

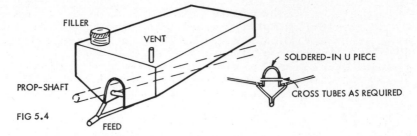

FIG 5.4

FILLER

VENT

PROP-SHAFT

FEED

SOLDERED-IN U PIECE

CROSS TUBES AS REQUIRED

In some boats the obvious site for the tank is immediately aft of the engine (it should be as close to the engine as possible) but the prop-shaft tube tends to be in the way. In such a case the tank can be modified by cutting a slot and soldering in a U-piece so that the tank sits astride the tube; tubes of reasonable diameter must be soldered across to allow all fuel to flow to one sump, or the fuel feed taken from the front cross-tube, or two feeds taken from the lowest point of each side and joined through a 'Y' connector for a single feed to the engine. Klunk tanks can be used, the feed being flexible inside the tank with a weighted end, so that the forces throwing the fuel from one side to the other also move the feed pick-up. Two such tanks can be used either side of the prop-tube, with twin feed to a Y-piece, but the tanks must be linked at the bottom; a neoprene tube connector between bushes in the tanks will achieve this.

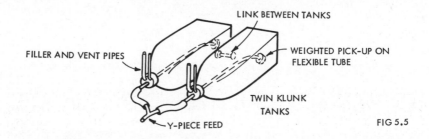

LINK BETWEEN TANKS

FILLER AND VENT PIPES

WEIGHTED PICK-UP ON FLEXIBLE TUBE

TWIN KLUNK TANKS

Y-PIECE FEED

FIG 5.5

If it is not possible to get the volume required low enough and the full fuel level is slightly above the engine spray-bar, a restrictor clamp on the feed pipe may prevent flooding when starting, the clamp being quickly removed as the engine starts.

Alignment

Alignment is one of the most vital parts of any power set-up, i.c. or electric. The motor shaft and the propeller shaft should ideally form one continuous line through the coupling so that no side loads are introduced. Some couplings allow a degree or two of angular difference, but even this is undesirable unless the coupling is of the type with two flexible joints. Side loads create wear on the motor and prop-shaft bearings, increase noise, and produce a braking effect which costs motor r.p.m. and can make an i.c. engine difficult or impossible to start, or flatten the batteries of an electric installation. One of the neatest ideas of recent years is an adapter that screws on to motor and prop-shafts, connecting them rigidly while the installation is carried out. The adapter is the same length exactly as the same maker's coupling, which is substituted when the mounting is completed, thus ensuring absolute precision of alignment.

The engine bay area should be thoroughly painted/varnished and/or fuel-proofed, and should be made easy to wipe out. Most engines rely on fuel seepage to lubricate the crankshaft and this fuel will emerge from the end bearing and be thrown off the flywheel. Wiping out after running prevents the accumulation of sticky sludge which can very well reduce the life of a boat.

5·7 *This diesel motor is designed for tuned pipe operation and has an unusual throttle operating on the exhaust outlet.*

5·8 *A multi-racer with a large part of the hull taken up by the fuel tank and an under-deck tuned pipe exhaust.*

59

Cooling

Water cooling is standard for i.c. engines (some high performance electric motors use it, too) and is nearly always confined to a water jacket round the cylinder head. A pick-up immediately behind the upward-moving side of the propeller is connected by neoprene tube to the jacket inlet and a tube from the outlet is carried to a bush on the hull side or transom. Sometimes the outlet is led into the exhaust pipe, cooling the exhaust fractionally and possibly making a marginal noise reduction. Generally, however, a side exit is best, so that the discharge of water can be visually checked while the boat is running. A common fault is over-cooling; the discharge should be hot and a small clamp on the outlet tube should be adjusted to reduce the flow until the emerging water is uncomfortable to the fingers.

Sail

Sails are a form of power which is rarely fully thought out. If all boats in a race are limited to similar sail area, since they are all using the same wind the most efficient suit of sails will produce the fastest boat, other things being roughly equal. Sails are airfoils, and their sections and shapes are as important as aircraft wings, which is why some expertise is required to make a good suit.

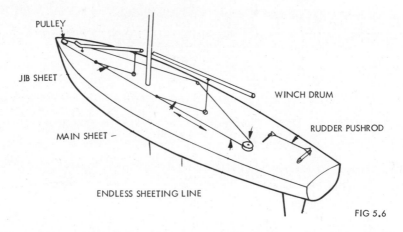

PULLEY

JIB SHEET

WINCH DRUM

RUDDER PUSHROD

MAIN SHEET —

ENDLESS SHEETING LINE

FIG 5.6

The usual material used nowadays is hot-rolled dinghy nylon, which can be obtained in various weights and bought by the metre from one of the specialist model sailmakers if an attempt at making sails is considered. Essential points are: that the leach (after edge) of all triangular sails must be parallel with the selvedge of the cloth and that, with synthetic cloth, synthetic tapes and threads must be used. The cloth is cut with a hot, sharp iron (a sharpened soldering iron will do), the luff (fore edge) is sewn into a folded tape and the leach and foot are left as cut, i.e. not hemmed. To introduce shape into the sail the luff is cut with a faint curve but sewn into the tape straight; where the maximum curve occurs influences the position of the flow or belly in the sail. It is recommended that potential sailmakers read fuller details in one of the specialist model yachting books available.

Present practice with racing yachts is to use the winch to operate a sheeting line laid along the deck, the line coming off the winch drum aft, round a pulley at the bow and back on to the other side of the drum. Thus rotation of the drum moves the line but does not affect its length, which avoids slackness and possibilities of tangling. The sheets from the booms pass through eyes on the deck centre line, or mounted on tripods on the centre line, and are made off to the sheeting line. Adjustment is provided on the booms to vary the sheet lengths if necessary. Both jib and main sheets must travel in the same direction and all travel should take place on the same side of the centre eye, allowing the sails to swing through 90 degrees. More than one novice has found that while one sail moves out the other moves halfway and then comes back in again, requiring his sheets to be repositioned.

This same system of a sheeting line moving along the deck can be used on scale-type models, for any number of sails (within the power of the winch), with the line perhaps run through eyes close to the bulwarks to be unobtrusive and the winch concealed beneath an item of deck furniture.

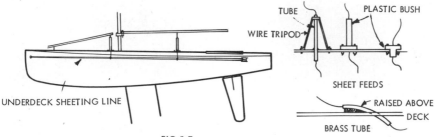

FIG 5.7

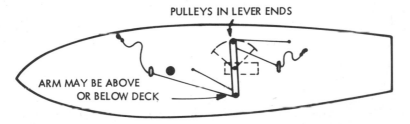

PULLEYS IN LEVER ENDS

ARM MAY BE ABOVE OR BELOW DECK

SHEET SYSTEM WITH LEVER WINCH

FIG 5.8

Lots of room in this large Range Safety Launch, which has electric auxiliary power. Note wiring tucked neatly under side decks.

Tighter fit in Glynn Guest's destroyer escort. Empty space amidships takes drive batteries.

CHAPTER SIX

The Installation

To begin with a fairly obvious fact, water and radio do not mix, especially if it is salt water. However, should radio equipment get a soaking, nine times out of ten prompt action can leave it no worse off. The wet items should be immediately unplugged and their cases removed, then everything thoroughly rinsed in clean water, preferably under a gently running tap. Excess water should be shaken or blown gently off and the equipment dried in gentle heat, such as above (not on) a hot tank in an airing cupboard, or in front of a fan heater. If the ducking was in salt water, this treatment should be carried out within a few minutes and rinsing must be very thorough. Some experts recommend a light spray with a water inhibitor such as WD40, as sold for car ignition system treatment, but usually rinsing and drying is adequate. Periodic checks should be made for some weeks afterwards to see that no corrosion is occurring, especially round soldered joints.

Receiver box

It is obviously better if equipment does not get wet, and to this end it is customary to enclose it in a box which may be splash-proof or waterproof. Splash-proofing means protection against water dripping off the hands or a cleaning swab, or against any small amount of water finding its way into the hull, and is adequate for most models. If any model will be run in very rough water more care may be needed, and fast competition models are best if the radio is totally watertight, since they may fill with water, flip or even sink.

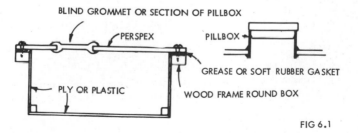

FIG 6.1

6·1 *Complete radio box for a yacht. Winch drum and switch wire are at top left.*

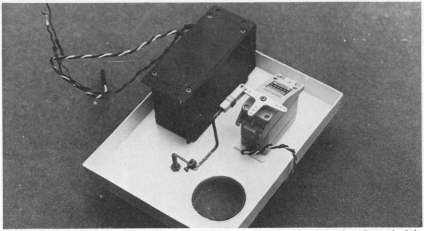

6·2 *The winch and rudder servo are mounted on adhesive pads on the underside on the lid.*

Some servos are waterproof as purchased, and this simplifies matters, since if they are mounted in a radio box, the push-rod exits are likely to the most awkward items to seal and the box has to be precisely positioned. With only the receiver and battery pack in a box, positioning is less critical and the box can be strapped in place with rubber bands, possibly on a bed of foam rubber. The foam is not as essential for running vibration as it was in the days of relays, when vibration could cause the relays to chatter, but it reduces the chance of a joint vibrating loose, either when the boat's engine is running or during transport to the lakeside by car.

6·3 Access to the crystal is via a pillbox top. Rudder operating arm is visible.

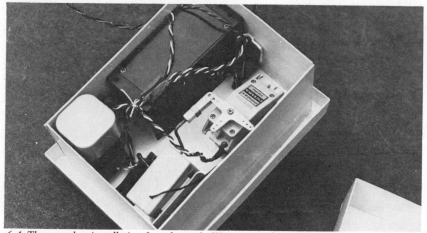

6·4 The complete installation from beneath. Wires are tied out of harm's way.

One difficulty is that, unless a special switch device is fitted, access to the receiver is needed to change the crystal. Rather than have to open the complete box, it is possible to mount the receiver in such a position as to bring the crystal under a circular hole which can be sealed with a large blind rubber grommet (from motorist shops), or the top of a pill-box etc. with a tight plastic cap, the cylindrical body being glued or epoxied into the hole. Enough room for fingers to grip the crystal is required, and there will usually be room to include the on/off switch under the same seal.

Non-waterproof servos can be given greater protection by running a line of bath sealant round the case joint(s) and putting a dab over the case assembly screwheads, or by wrapping the case completely with plastic insulation tape. Bath sealant will peel away if necessary, leaving the surface unmarked; an alternative is a latex adhesive. A generous smear of silicon grease round the output shaft, beneath the disc, will deter water from working down the shaft. Where a disc appears above deck (not uncommon on yachts) a thin ring of felt smeared with grease, beneath the disc, will not impose undue load but will prevent water weeping through the deck or down the servo output shaft.

Constructing the receiver box

Plastic food storage boxes are often used to contain the radio but often need stiffening with an internal ply shelf and usually suffer from opacity. It is convenient to be able to inspect the installation without removing the lid, so it is better to make a box from thin ply, with a fairly substantial wood strip frame in or around the top. A Perspex or other stiff acrylic cover can then be cut to size, drilled about every 2in. (50mm), laid in place on the box and the holes spotted through on the frame. Smaller holes are then drilled in the frame so that the cover can be screwed in place. The cover requires a hole as previously described. After painting or varnishing the box inside and out and installing the radio, the cover can be screwed in place. A film of silicon grease, a soft rubber gasket, or strips or draught-proofing foam can be used between frame and cover, and can be seen to spread evenly as the screws are tightened. If foam is used, it needs to be well compressed or water could soak through it.

A development of this, especially for small boats, is to seal off the entire radio compartment from the rest of the hull, by painting resin round the bulkhead joints etc., and make the frame the hatch outline, raised slightly above deck and sanded absolutely flat. The screwed-on cover then forms the hatch itself.

Any sealed box like this should have inside a small bag of silica gel. The crystals can be bought from a chemist and small bags made from cotton or linen, perhaps the sound parts of an old handkerchief. These are filled with the crystals and sewn

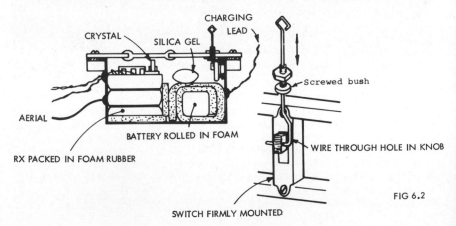

FIG 6.2

up. This substance absorbs free moisture in the air, and will thus keep the atmosphere in the sealed box dry. It is necessary to change the bags from time to time, drying out the used ones in an airing cupboard or near a radiator ready for re-use.

If the box is only for receiver and battery pack, they can be rolled in foam rubber (preferable to foam plastic) and tucked in, with servo, charging leads and aerial passing out through small holes sealed with bath sealant. If it is desired to fit the Rx positively for crystal changing, it can be mounted on double-sided adhesive foam pads, which will stick firmly to any smooth surface, hence one reason for painting or varnishing the inside of the box. Depending on the Rx and the crystal socket position, it may be possible to stick the Rx to the underside of the acrylic cover. The sticky pads absorb vibration and will allow removal of equipment, even if by actually tearing the foam core and then peeling off the adhesive patches.

Switches mounted to be operated externally can let water in, or may corrode, which is why it is suggested that they are mounted under the crystal-changing hatch. An alternative is to mount the switch firmly inside the box and use a wire through a close-fitting bush in the top or side of the box to operate the switch. A little grease on the wire guards against water, and it is not difficult to cross-drill the switch knob to receive the wire; some switches are provided with such a hole ready-drilled.

6·5 *Receiver and batteries are in a waterproof plastic box inside the hull, but the servos in this instance are waterproof and mounted at deck level*

Wiring leads to servos should be anchored near each end, so that any accidental pull is not transmitted to the plug or servo connections. To reduce the chance of this happening, leads should be tucked neatly out of the way, held in position by insulation tape, staples made from bent pins, or thread bindings to appropriate parts of the structure or screw-eyes positioned for the purpose. Above all, ensure that no wires can foul the servo linkages, especially if they are mounted inside the same box as the Rx and battery. In this case, coil the leads and tape them positively out of the way.

If servos are to be mounted inside the box, a little more is called for. Either the base of the box must be thick enough to receive screws, for foot-mounted servos or the special servo mounts which are available for most makes, or rails must be provided on which to mount the servos. Rails (of ¼ × ⅜in., say 6 × 9mm wood) are really better than cut-outs in a ply shelf, provided they can be glued firmly in place and that they do not obstruct access to the rest of the box. Small rubber grommets are supplied with servos and screws passed through them should just pinch the grommets, not flatten them. If a ply shelf is used, bolts instead of woodscrews may be needed, and these should be tightened in the same way. A touch of Loctite or epoxy will then be necessary on the threads to ensure that the nuts cannot work loose.

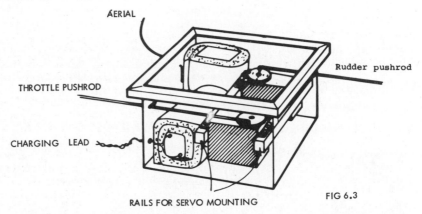

FIG 6.3

Such a box will require small holes for the aerial and the charging lead (a lead is easier to waterproof than a charging socket) and it will also need clearances provided for the servo linkages.

Throttle and rudder linkage

Linear servos, where the output slides back and forth in a straight line, offer no problem, since the pushrod can pass through a close-fitting bush which, with a smear of silicon grease, will be reasonably watertight. The more normal rotary output servo, used with a conventional wire pushrod, suffers from sideways displacement as the disc rotates, so that a slot is needed to pass the pushrod. This may be accentuated in the case of a rudder mounted close to the radio box, where the tiller also moves the pushrod sideways.

A standard cure is to cut the neck from a rubber balloon and trap the wide end against the outside of the box with a large fibre (or similar) washer with a cut-out centre to clear the pushrod slot. Depending on the stiffness of the washer, three or four small bolts and nuts are needed. The other end of the piece of balloon is wrapped and bound to the pushrod, forming a watertight gaiter over it which will not affect its movement. Soft rubber 'top hat' bellows are commercially available, complete with fixing washers, to do the same job.

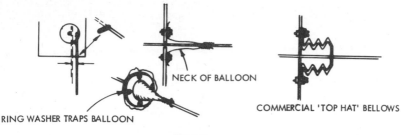

RING WASHER TRAPS BALLOON

NECK OF BALLOON

COMMERCIAL 'TOP HAT' BELLOWS

FIG 6.4

An alternative is to use a 'Z' crank through a bush in the side or top of the box. The servo is connected by a short link to the inner arm of the 'Z' and the outer arm is then connected to the tiller or throttle arm etc. There is only one minor drawback, and that is the chance of introducing slop (free movement) into the linkage, but with care this can be minimised.

Another possibility is to use 'snakes', the type of pushrod consisting of a stranded wire or thin nylon tube sliding inside an outer casing. It is necessary to mount the servo two or three inches from the exit point in the box side, to allow sufficient sideways movement of the inner core of the snake; the casing must also be positioned firmly by clamping lightly to the hull structure at three- or four-inch intervals, using as straight a run as possible. Snakes are often convenient for throttle control, but are less useful for rudders. Ideally, the casing should be immovable, or operation of the inner core can increase or decrease bend radii, leading to sloppy control.

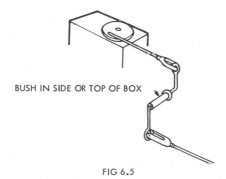

BUSH IN SIDE OR TOP OF BOX

FIG 6.5

One way to reduce linkage problems and often mount the radio box in a convenient spot is to build it into the boat with the rudder stock (shaft) actually inside the box, when a simple link to the servo can be fitted. To prevent water working up the rudder trunk a bushed tube can be used, with a small tube soldered into it leading to a grease nipple on the transom, or on deck. The rudder tube is pumped full of grease and an occasional squirt given during running to keep it topped up and water-free.

69

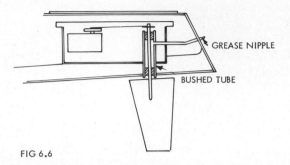

FIG 6.6

Wire pushrods for throttle linkages should again be kept as straight as possible, as every bend introduces spring and absorbs movement. A plain pushrod is really only suitable for front rotary induction engines or other installations where the throttle lever is on the aft side of the engine as installed in the boat; there is bound to be a little confusion of terms since most existing engines are marinised versions of aircraft motors and retain their terminology, so that a rear induction engine actually has its intake ahead of the engine in a boat! Where the throttle is aft of the engine a pushrod can be fitted on or near the centre-line of the hull and, surprisingly to beginners, is not likely to get in the way of the starting cord. If a large metal tank is sited between servo and engine, it is possible to solder a tube through it, but if plastic is used, it is better to fit twin-linked tanks as described, separated fractionally to allow passage of the pushrod.

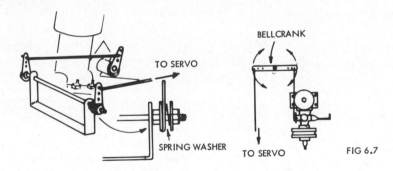

FIG 6.7

Where the throttle lever is ahead of the engine, or positioned well off-centre, it is better to introduce a directional change in the linkage, either by fitting a bell-crank or, better, by fitting a pivot rod. This is rotated by the movement of the pushrod via a horn at one end, and a second horn mounted at a convenient point along the rod is linked direct to the throttle lever. If the end horn, or the pivot of the bell-crank, is a threaded arrangement, the horn or crank can be gripped between two nuts with a spring washer introduced. This will allow the horn or crank to slip when it meets firm resistance, which has two benefits. Firstly, the

6·6 *In this example the throttle servo link (a tube for stiffness) passes just clear of the space needed for the starting cord.*

6·7 *A plastic clevis on the servo pushrod engaged with a clamp-type plastic tiller means no chance of metal-to-metal interference.*

pushrod travel can be arranged to exceed the throttle lever movement, thus ensuring that the throttle always moves the full distance, and secondly the throttle can be adjusted manually for starting and the linkage will automatically adjust immediately the throttle servo is operated in the same sense as the manual adjustment.

Wire pushrods can be made from any suitable and preferably rustproof stiff wire of 16swg. (1.5mm) upward (cycle spokes are useful), or they can be bought commercially produced. The commercial variety have one or both ends threaded and fitted with a clevis, which is a long fork of springy metal or plastic, one side of which carries a peg to engage in the servo disc etc., being retained in place by the tendency of the fork sides to spring together. Fine adjustment to lengths can be made by screwing or unscrewing one or both clevises.

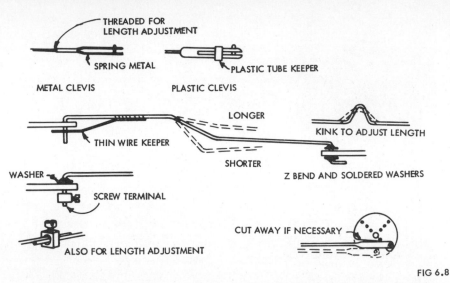

THREADED FOR LENGTH ADJUSTMENT

SPRING METAL

METAL CLEVIS

PLASTIC TUBE KEEPER

PLASTIC CLEVIS

LONGER

THIN WIRE KEEPER

SHORTER

KINK TO ADJUST LENGTH

Z BEND AND SOLDERED WASHERS

WASHER

SCREW TERMINAL

ALSO FOR LENGTH ADJUSTMENT

CUT AWAY IF NECESSARY

FIG 6.8

Home-made pushrods can achieve a similar effect by bending the end of the wire to a right-angle to enage the disc or control and fitting a thinner springy wire keeper, bound and soldered to the main wire as drawn. Kinking the pushrod, and opening or closing the kink, provides fractional length alteration. Alternatively the ends can be bent to a 'Z' shape, worked through the appropriate holes and fitted with a washer soldered to each side, loose enough to ensure free movement. It is easiest if the disc is removed from the servo for this, but care should be taken not to use so much heat as to melt the disc. One way to avoid this is to solder one washer to the wire and use a screwed fitting on the other side, with a loose washer; suitable screwed items can be cut from the flex-retaining parts of damaged electric light fittings and plugs. Another use for these parts is to provide length adjustment by having two shorter pushrods overlapping through a screwed block, the screw being slackened to adjust the relative positions.

It is essential that all linkages be freely moving, since any friction or binding will flatten batteries more rapidly and may even damage the servo. Check over the whole range of movement, and cut away any item which obstructs full and free movement. This may include part of the servo disc or the tiller etc. At the same time, avoid slop or rattle such as may be caused by a pushrod engaging in too large a hole.

When planning an installation, check that the servo movement will give correctly sensed responses, i.e. left rudder on the Tx actually gives left rudder on the boat and that 'throttle open' means that, especially where a bell-crank etc. may be fitted in the linkage. It is possible to change to the other side of the disc (or reverse the disc) but this can produce unfavourable geometry, e.g. very little right rudder but excessive left rudder. In extremes it may be possible to fit a servo reverser, or change the servo to one of opposite sense, but in many cases the servo can be moved to a position where the linkage does line up. Another pitfall is fitting the linkage with the servo disc not in a neutral position, which is not difficult to do with non-sprung Tx sticks. A thorough check beforehand, with the radio working, will avoid such problems.

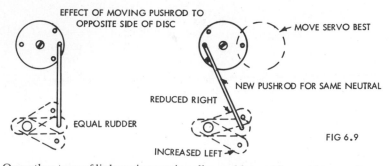

EFFECT OF MOVING PUSHROD TO
OPPOSITE SIDE OF DISC

MOVE SERVO BEST

NEW PUSHROD FOR SAME NEUTRAL

REDUCED RIGHT

EQUAL RUDDER

INCREASED LEFT

FIG 6.9

One other type of linkage is occasionally used for rudders and is usually termed a closed loop. In this a light wire or even braided Terylene line is used to connect both sides of the servo disc or arm to a tiller extending each side of the rudder stock, so that rudder is always applied by a pull rather than a pull one way and a push the other. Its main advantages are very precise and positive rudder movement and the opportunity of taking the line round corners without introducing differential rudder throw, provided, of course, that too much friction is not introduced. A useful variation might be scale-type rudder operation with a forward pointing tiller, both lines being attached to the end of the tiller as on a tug or puffer etc. The lines can change direction over polished wire guides or small free-running pulleys, or through ceramic fishing rod rings.

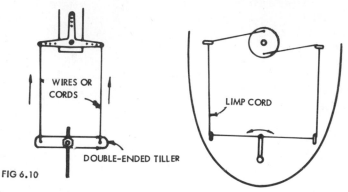

WIRES OR
CORDS

LIMP CORD

DOUBLE-ENDED TILLER

FIG 6.10

Rudder movement

Rudder movement in full-size does not often exceed 25 degrees either side of centre, but is frequently more than this with models. Care should be taken that the rudder cannot turn so far as to be forced 'over centre' by the propeller wash, as can happen with a balanced rudder and a springy pushrod when excessive rudder angle is used. To avoid spring in the system, anchor the rudder tube firmly with blocks or gussets and, if a long pushrod is essential, use a ³⁄₁₆in. dowel, or aluminium tube, with wire ends bound and epoxied in place. Many long, narrow scale models with electric motors and twin or quadruple screws are very slow to turn, and it is worth considering installing microswitches operated by the tiller at full rudder to break, or even reverse, the current flow to the inside motor of the turn.

On fast boats, rudder shape and distance from propeller are important factors, but they vary with engine r.p.m., propeller pitch, hull shape and speed. Equal rudder movement may not produce equal diameter turns in either direction, due to the manner in which the high-pressure water streams from the propeller blades impinge on the rudder surface, giving it more bite one way than the other. The cure is to change the rudder shape by cropping diagonally opposite corners (front top and bottom rear, or front bottom and top rear) and there is a 50/50 chance of being right first time! Sometimes rudder bias is needed to offset the turning effect of propeller torque and achieve straight running, but usually this can be introduced with the rudder trim on the transmitter.

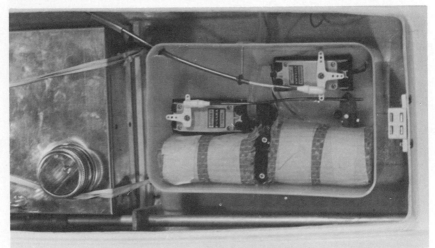

6·8 *In this installation the tiller is inside the radio box and a 'snake' pushrod is used for the throttle*

Most servo discs move through 45° (sometimes up to 55°) each side of neutral, and to achieve the same degree of rudder movement the tiller radius must be the same as the disc radius. Extended arms are available for most servos, allowing a correspondingly longer tiller and therefore reducing free movement; the same degree of fit can be achieved, but since the movement is greater, any free movement (say from a wire pushrod loose in the holes) will be a smaller proportion of the total. Since 45° of rudder is rarely needed, the tiller can be made about 1½ times disc radius, which will give some 30° of rudder movement each side, normally an adequate amount. By ensuring that the pushrod meets the tiller at right-angles when centred, an equal amount of rudder each side is assured.

Metal-to-metal joints are better avoided, since they can produce 'noise interference' with the radio, but if they are used and interference seems possible, a short piece of light hook-up wire can be coiled and the ends soldered to pushrod and control (e.g. between pushrod and tiller) so that a permanent connection is made without hindering free movement. In many cases a plastic clevis can be

74

used, or a plastic or other non-conductive material used for the tiller etc., when the problem should not arise.

Aerials

Manufacturers supply receivers with a length of limp plastic-covered wire as an aerial, and the length supplied is important. The instructions may also specify a definite length to be used if the aerial supplied is changed. This length forms a part of the tuning of the receiver for maximum sensitivity and any substantial alteration will reduce range. As it happens, range is not usually crucial on boats and a reduction in length is therefore acceptable; the ideal length is often about 26in. (660mm.) from receiver to aerial tip, but cutting back to about 18in. (460mm.) is common practice.

The Tx aerial is usually nearer vertical than horizontal, and to make the best use of a shortened Rx aerial this too should be near vertical, hence the widespread use of piano wire whip aerials on boats. Running the aerial along inside the boat or along the deck shortens range, though for scale models not likely to venture more than 70-80yds. from the Tx this is often adequate. Passing the aerial near an electric motor (even in the servos) may give interference, but only a trial will show if this occurs at a significant level. Coiling the aerial, or doubling it back on itself, is effectively shortening it to the distance between the receiver and the furthest point of the coil or doubled wire, and while this may be adequate on a swimming pool, it is a dubious practice on a larger lake, or where other models are operating.

If a plug-in whip aerial is used, or if the limp wire from the receiver is joined to, say, a wire shroud or mast stay, to act as an aerial, good contact at the joint is essential. A poor joint will reduce the strength of the radio signal reaching the Rx, thus reducing range. If a telescopic aerial is fitted, it should be kept in a clean and sound condition for the same reason. Probably the best answer is a permanently fitted thin piano wire-whip aerial which can be bowed forward and hooked down

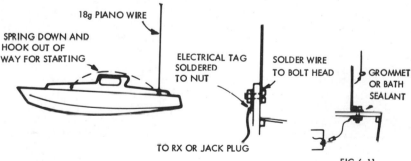

FIG 6.11

to be out of the way for starting; it will still work well if the boat is inadvertently launched without unhooking it, at least enough to bring the boat back in to release it. If the wire is bent into a ring at the top, or fitted with a plastic button, the chance of accidental face-scratching will be reduced.

Where a disguised aerial is used on a scale model, a check on total length should be made and an insulator introduced if it exceeds the recommended aerial length. This could also happen if, say, the wire backstay of a Marblehead yacht were to used as an aerial, though a better solution is to run the makers' aerial up the stay and tape it in place, but only if the stay is of cord.

A useful tip for those satisfied with running the aerial along inside the hull is to thread it through drinking straws or a plastic tube from a reed blind which will ensure that it is fully extended through inaccessible parts of the hull. An alternative is to tape or rubber-band the end to a length of balsa so that it can be fed right into the bow of, for example, a GRP yacht hull, most of which is permanently covered by the deck with only a small hatch well aft.

6·9 *The interior of a paddle-steamer model using dry batteries for power. Narrow hulls can make installations crowded.*

Fitting the receiver box

The radio box may be glued or screwed into the hull or it may be removable, perhaps for storage indoors, maintenance, or transfer to another model, in which case it can be positioned by a frame built into the hull (possibly with foam packing) and held firmly by strong rubber bands hooked over brass cuphooks screwed into the structure or into blocks specially provided for them. It may only be necessary to disengage the pushrods and unplug the aerial, or, if the servos are externally mounted, unplug them.

Finally, where a box is for some reason undesirable or inconvenient, never overlook plastic bags. A receiver or battery pack can be slipped into a bag, the neck twisted up and a rubber band wound tightly to hold the bag to the emerging wire and the package will be virtually waterproof. Undo the bags after a day's sailing unless a packet of silica gel has been included.

6·10 A transparent hatch enables a visual check on conditions inside this RM radio box

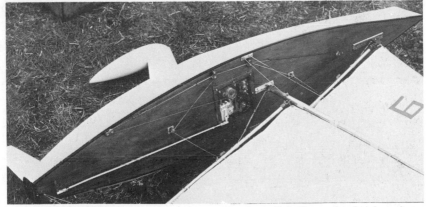

6·11 In contrast with the yacht above, this one has above-deck tiller pushrod and winch drum.

77

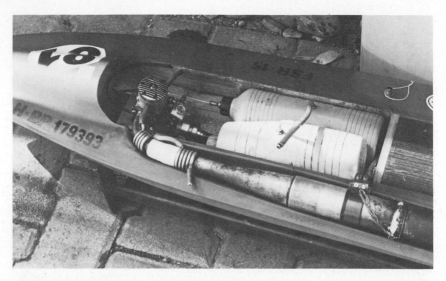

Typical modern ·60 powered multi-racer. Fuel tanks are twin linked plastic bottles and outlet from water jacket is used to cool exhaust header and 'hot' end of tuned pipe. Wire at right aerial lead or possible earth.

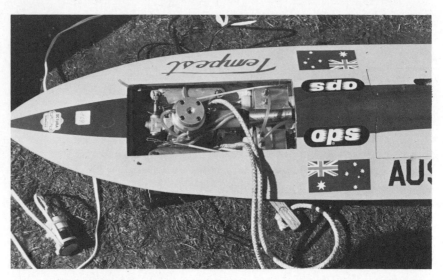

This Australian racing model uses twin linked metal tanks and flexible motor mount. Angled throttle push-rod clear of starting cord can be seen. Central exhaust/tuned pipe does not interfere with starting procedure.

CHAPTER SEVEN

Checks and Operation

There is an enormaous temptation, once the radio is installed, to rush off to the water and try it all out, but before doing so a complete check over the entire model and installation is necessary. This, in the case of a power boat, means seeing that the final tightening down has not pulled the motor/prop shaft out of alignment and that with the propeller in place the whole drive turns quite freely. Unless a glow engine etc. is new and tight (and it should have been run in for a few minutes on the bench to check starting and running) it should be possible to flick the propeller and 'bounce' the piston against compression. An electric motor should be felt to be 'lumpy' as the armature is turned through the magnetic field by rotating the propeller with finger and thumb.

Boat checks

Check for free movement of the rudder – paint has a habit of working up the shaft and should have been scraped off – and ensure that it is vertical in the hull and that the tiller is secure; pinch-bolt-type plastic tillers on smooth shafts, or metal/plastic ones nutted on to the threaded shafts, have a habit of slipping unless firmly tightened. There are times when this may be an advantage, if the boat is laid carelessly on the ground and the rudder banged, but on a stand the rudder should not easily be knocked. You did make a stand? Make sure that hatches fit and that any clips, catches, rubber bands etc. are in working order, that the fuel tank is firmly mounted and the fuel and water tubes are properly engaged. A loose water inlet pipe can fill a boat with water surprisingly quickly, so if there is any doubt,

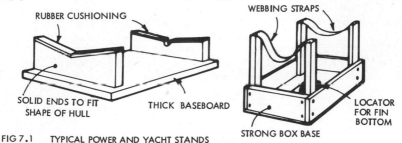

RUBBER CUSHIONING
SOLID ENDS TO FIT SHAPE OF HULL
THICK BASEBOARD

WEBBING STRAPS
LOCATOR FOR FIN BOTTOM
STRONG BOX BASE

FIG 7.1   TYPICAL POWER AND YACHT STANDS

79

use a twist of soft wire (twisted with pliers) to hold the tube, rather as a jubilee clip is used for hoses in car engines.

With a yacht, the essentials are fin and rudder truly vertical and accurately aligned fore and aft and, looking from ahead or astern, mast absolutely vertical, i.e. not leaning to one side. A check over all rigging hooks and eyes, knots in cord rigging, freedom of movement of booms, adequate room for adjustments where bowsies are used, and rudder and hatch(es) as mentioned above may well save fumbling about with inadequate facilities at the lakeside.

7·1 There was more to check in 1955, when this photograph was taken. Modern equipment is more reliable and easier to maintain.

80

Radio checks

Most radio troubles are likely to be traceable to batteries, switches, or broken wires in the wiring harness. Assuming that the equipment is new or little used, there should be no problem about its working when installed. If it doesn't, the checks in the next chapter can be applied. During construction and installation, the radio will have been on from time to time, enough to have affected the batteries, so if dry batteries are being used, get a new set. A characteristic of dry cells is that they will always deliver a slightly higher current when first switched on, settling down after half a minute or so; with earlier equipment, it has been known for new batteries to cause a small amount of servo interaction or twitching for a minute or so, before the cells settled down. With old batteries, there is the danger that a rest will have allowed sufficient recovery for them to operate the radio when first switched on, only for the current to fall to below the necessary minimum after a minute or two.

Rechargeable cells should always be charged fully before setting off to the water. At one time it was believed that continual charging for periods in excess of the actual time used would cause cell deterioration, but nowadays makers recommend a full charge (usually fourteen hours with the charger supplied with the equipment) irrespective of when the equipment was last used or for how long.

Check that all plugs are correctly and fully engaged with their sockets and that matching crystals are fitted, with the Tx one in the Tx and the Rx one in the Rx. Look over servo mountings and linkages (no binding or fouling, but no rattles) and sealing of the radio box, and always, always check that switches are off. The procedure is invariably Tx on, then Rx on, Rx off, then Tx off. Establish this, and a further check, as an ingrained habit; there have been many occasions when a competitor has arrived at a meeting only to find his batteries flat because he had failed to switch off after his final check the evening before.

Modern sets very rarely give any trouble, but a range check is a worthwhile exercise the first time out, or as second best it could be done at home. Coil up the Rx aerial, or disconnect it if attached to a whip, and prop the boat so that you can see the rudder. Leave the Tx aerial telescoped down, switch on Tx and Rx, walk away and operate the rudder. If at 10m it still operates, you should have no difficulty. If at the waterside, check with the motor running while an assistant holds the boat in the water. Reconnect or lay out the Rx aerial after switching off, and remember always to pull out the Tx aerial to its full extent when operating. Make sure, too, that it carries the correct colour pennant or ribbon for the frequency you are using.

You can expect to operate for about three hours before needing recharging or battery replacement, though this will obviously vary with the frequency with which rudder and throttle or sails are altered. Using a sail winch running off the Rx batteries may cut it to little over an hour; it is customary to use a battery of the same voltage but greater a/h capacity for yachts of 36in. and upward.

Most Txs nowadays have a meter indicating battery state, but it is usually the Rx power which fades first. With dry batteries the symptoms are often slower reactions of the servos, but with nicads the first sign is usually erratic operation of the servos. It is safest not to get to this stage, but at the slightest interruption to smooth control the model should be brought in.

If a model fails to respond to a signal, sometimes holding the transmitter above the head and moving it around will sufficiently increase received signal strength to save the situation. The alternative is to reduce the range by moving towards the model. In any event, when the model is retrieved carry out a range check as above, and repeat with fresh batteries, or after the nicads have been recharged, to ensure that the problem was caused by the batteries going down.

Cleaning

At the end of a session, on reaching home clean the boat by wiping with a light detergent solution or, if used in salt water, by hosing it off with clean fresh water. Check that no water has found its way inside, but if it has, wipe and dry out straight away. Remove all dry cells from equipment; it is fatally easy to leave them in and find circumstances preclude use for a few weeks, when any cell leakage will have caused corrosion. Should this happen, treat the corroded area with a paste made from bicarbonate of soda and water and wipe off thoroughly after a few minutes. Afterwards regular checks will be necessary to see that any electrical contacts remain clean and bright. If only the battery holder is affected, throw it away and replace with a new one.

7·2 *Checking models during a regatta. There is always someone in a club willing to help with problems.*

Spares

It pays to have a special small box for radio parts to be taken to the lake, to hold spare crystals (don't let them rattle around) and pennants, spare batteries, plastic insulation tape, paper tissues, rubber bands, even spare servo discs, plus a miniature cross-point screwdriver and a small electrical screwdriver. Cleanliness and tidiness make life easier and may help to prolong the life of the equipment, so don't throw all the radio bits in the bottom of the box with the fuel cans, glow accumulator, spanners etc.

Keeping the transmitter dry

In thinking about keeping the equipment dry, remember the transmitter. Competitions carry on even if it starts to rain, so keep a large clear polythene bag with the spares and slip this over the Tx, poking the aerial through a hole, in wet weather. The hands can be slipped into the bag and the controls operated under cover without difficulty. It is possible to buy purpose-made Tx muffs which perform this function, but a simple bag is adequate. Naturally an inspection for stray water drops is necessary on completion of the run, but a quick wipe with a clean tissue is likely to be all that is needed.

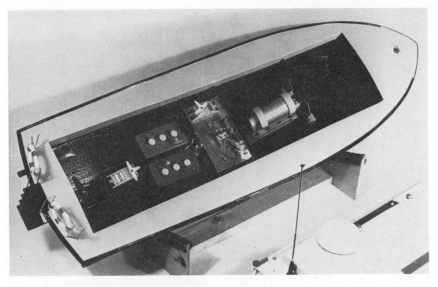

This pilot launch has lead/acid accumulators and so will not be very fast. Even so, the exposed speed control mechanism is vulnerable to damp and drips and will require regular checks for corrosion on contacts and wire joints.

A simple unit makes transfer of equipment between models simply a matter of disconnecting control push-rods and facilitates regular inspection and maintenance.

CHAPTER EIGHT

Maintenance and Fault-Finding

Modern equipment requires minimal maintenance, and cleanliness is the first step. Regular use is bound to show, particularly with internal combustion-engined boating, where some fuel on the hands is virtually unavoidable. An occasional wipe over the cases with a cloth lightly dampened with detergent will keep them clean; stubborn marks, or remnants of adhesive from insulation tape or sticky pads, will disapear if a rag damped with petrol (gasoline) is rubbed over them. Avoid using too much.

Maintenance

Battery cases can be cleaned with petrol or methylated spirits (alcohol) and contacts should be checked for cleanliness. Dry batteries should be inspected for bulges in the cover, cracks or other physical damage, and thrown away if in the least bit suspect. A voltmeter is a very desirable addition to the toolkit (a very simple and inexpensive one is sufficient) and the batteries can then be checked under working conditions, i.e. on load. If the meter probes are held on the Rx battery case terminals while an assistant operates the Tx to switch in a servo, any significant dip in voltage can readily be seen, and if there is a big dip, discard the cells. Better safe than sorry!

Nicads read 1.1v (each cell) when discharged, against 1.2v when fully charged, and a voltage reading will not indicate the state of charge, only whether the cell is discharged or has some degree of charge remaining. A really careful user will discharge the cells to a 1.1v reading, using a motorcycle or car bulb for the purpose, and then give a full charge. It is possible to buy cell recyclers which do this automatically, so that it is certain that the nicads will stay in tip-top condition and always be fully charged at the start of a boating session, but most people simply give a full fourteen-hour charge the day before use, irrespective of whether the cells already hold some charge.

It is desirable to run the cells right down, (recharge, discharge and recharge two or three times) at least once, but not more than twice, each year, which has

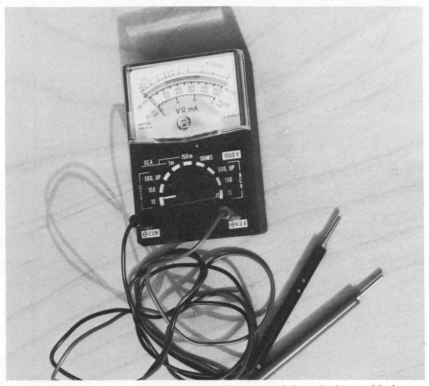

8·1 An inexpensive meter (this one cost about £5) is a great help in checking and fault-finding, but there are alternative methods.

beneficial effects. You will hear tales about 'reverse polarity' and 'memory effect' but most of these are unfounded and the occasional two- or three-full charge/discharge cycles will inhibit any internal changes which might lead to reduced cell performance.

Nicads, or Deacs as they are still occasionally called, should be kept clean and the terminal joints examined periodically. Should it be necessary to solder or resolder a tag, use a large and hot iron for as brief a period as possible to ensure a sound joint, since heat may damage the cells. Button cells (the type most often referred to as Deacs) are usually supplied in packs enclosed in a thick shrunk-on plastic tube; should the tube be sufficiently damaged as to threaten possible loosening of the pack, it should be returned to the makers for re-sleeving.

Working from the Rx battery, examine all the wiring harness for kinks, cuts, broken or frayed strands or connectors, or loose nuts etc. on threaded terminals. Unless you are experienced with a soldering iron, return anything suspect to the manufacturer or his service agent. The same goes for the switch, often a source of trouble. At the slightest sign of it not functioning perfectly, change it or have it changed. If it is simply a dirty contact, possibly cleaning with a special cleaner and lubricant (available from radio shops) will be adequate, but if there is any doubt it is cheap insurance to see that it is replaced.

There is little to be done with receiver and servos, mainly a check for defects in any loose wires or leads (including the aerial) and an inspection for physical damage of any sort. It is not even recommended that the servo case should be removed to examine the gear train unless operation has been suspect in any way. Servos requiring lubrication are supplied with instructions by the manufacturer and little difficulty should arise if the instructions are faithfully followed.

Maintenance, in other words, is a matter of common sense allied with recognition of one's own limitations and a refusal to yield to the temptation of taking things apart unnecessarily. Today's equipment will give many hours of use – at least two or three seasons – before any repair becomes necessary, provided it is not abused. If it is heavily used every week-end, it is a good idea to return it to the maker or agent for an annual check, which should not only avoid a possible breakdown at the height of the season but will prolong the life of the set and very probably save money in the long run.

Fault-finding

As with any electronic equipment, there is always the chance that one day something will fail to work. Usually the reason is traceable to something simple like a broken wire which cannot easily be detected, but occasionally a component might fail. Being able to locate the fault is the first step, and this should start with a check on obvious things such as the correct pair of crystals correctly positioned in the Tx and Rx and the aerial and battery connections sound. To go further, the biggest helps are a friend or clubmate with a similar set, and a means of checking simple continuity of circuits, such as a voltmeter or a 1½v battery wired to a 1½v bulb.

Elimination procedure is perhaps best demonstrated in a simple situation where, say, one servo fails to operate although the other does. Plugging the dead servo into the working one's socket will show whether the fault is with the servo or the receiver connection to the socket; the latter can be checked by plugging the working servo into the apparently dead socket. If it doesn't work and there is no obvious loose wire in the Rx, the Rx will have to be sent for repair.

Intermittent operation calls for a check on aerial and battery connections and all other wiring; sometimes a wire broken inside its insulation will make erratic contact. Check the on/off switch in particular. Sometimes an internal battery fault will result in a fluctuating current, which needs a voltmeter for a check. Twitching of the servos may occur on fully-charged batteries and the Tx held too close, or sparking from the motor of a well-used servo may cause intcraction. Interference from another set may cause glitches (unexpected servo movements or other unusual behaviour) or there may be other interference (dealt with separately). If none of these appears to be the cause, send the whole set for checking.

One servo intermittent can be checked as above, or plugged into a friend's identical receiver. If it is not the lead wires, it is possibly the feedback potentiometer; some servos use open types (most are sealed) and the track and/or wiper could require cleaning with contact fluid/lubricant. Otherwise, it's back to the makers.

All the equipment totally dead can be more of a problem. Switching on the receiver by itself should cause an instantaneous twitch of the servos, and though this doesn't necessarily mean it should work, it indicates that there is some life and that therefore the Tx may not be functioning. If the battery is O.K., check the aerial connection. A simple check on whether the Tx is in fact transmitting is to switch on and hold the aerial very near the aerial of a portable radio, when a buzzing noise should be audible. Movement of the control stick will alter the character of the buzz. A further check is to see if it will operate a friend's Rx, with matching crystals, of course. If not, return Tx and Rx for checking.

Assuming that the Tx is found to be working, check the Rx battery and then the battery input to the Rx, which will show up a faulty switch. If the current is reaching the Rx and there are no obviously displaced wires, it's off to the makers with it.

Having a second set available means that Tx, Rx and servos can be interchanged one at a time and the fault narrowed down to one part. However, if there is nothing obviously wrong it is best to send the whole set to the makers or service agent; service is normally very quick, especially if you explain any urgency. Pack the equipment carefully and include a letter explaining exactly what happened, and whether the equipment has been dropped or wetted. The experts will find the fault very much quicker than a radio enthusiast unfamiliar with the type of equipment, and the charge for repair will be fair.

Interference

Development, both in circuitry and components, has very much reduced the sensitivity of modern radio equipment to interference, but should it arise, it is as well to know what steps can be taken. It can be caused by transmission on an overlapping frequency, harmonics from other transmissions, or spark generation, and unfortunately in the first two cases there is not a great deal to be done other than try a different set of crystals. There have been cases of some clubs operating near a source of transmission where it has been found that one frequency can never be used, although no trouble is found on any of the others. Much industrial or medical equipment likely to emit interfering signals has very limited range, and harmonics from more powerful transmitters may occur only at limited times or only under unusual meteorological conditions. However, a major problem is the

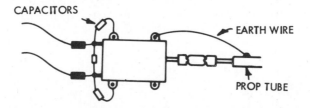

spread of Citizens' Band radio. The difficulty is that much of the equipment used, illegally at present but likely to continue in use even after a legal waveband has been allocated, directly interferes with model control on any or all model frequencies. Persistent difficulty should be raised with the local telephone manager, who is empowered to look into the matter. It is worth trying reducing the Rx aerial to reduce sensitivity while still retaining enough range for reasonable operation.

The range of transmission of a small spark is limited to a few feet, but it is not possible to mount the radio a few feet from the source of sparks if it is in a boat with a spark-ignition motor or an electric motor sparking at the brushes. Obviously, however, it should be sited as far away as possible and the aerial kept well away from the source. It may also help to keep all the wiring harness clear of the sparking area. Sometimes a metal screen between motor and radio helps, and this might be as simple as covering a bulkhead with kitchen foil.

Reducing, suppressing, or locally screening the spark are the best measures. In the case of a spark-ignition engine, reduction of the spark at the plug is impractical, but a screened plug lead and suppressor cap will normally suffice, just the same sort of fittings which reduce TV interference from cars and mowers, in fact.

Sparking can be reduced in electric motors by cleaning the commutator and ensuring that it is smooth and true, by cleaning the brushes and bedding them in properly, and by applying a trace of Electrolube or a similar contact cleaner/ lubricant. Additional steps are to 'earth' the motor case to the propeller shaft tube with a length of wire soldered in place or, where a plastic case is used, to a tag attached to or even rubbing on a metal part of the motor, and to fit capacitors and/or chokes. Chokes in the battery leads, close to the motor terminals, should be between five and eight microhenries; capacitors, of about .01 microfarad (up to .05mfd may be needed), should be fitted between each brush and the motor frame, and/or between the brushes. The same treatment can be applied to any electric motor, including those in the servos if necessary.

With modern radio, it would be surprising if any of these measures were necessary, but it is always possible that a heavily-sparking motor may be encountered. Usually, however, today's equipment is strictly plug in and switch on; with only a little care it should give pleasure for years.

A *A fine model of a Clyde 'puffer', effectively a steam-powered lighter. Scale models are now the most numerous of all.*

B *Fast and exciting models are rarely exactly to scale but this one is close to a Fairey Swordsman.*

C *Neatly constructed RM yacht with a minimum of clutter on deck. There is more to sailing than might appear.*

D *Scale model of* Bismarck, *an example of some of the superb kits which have been developed by European manufacturers.*

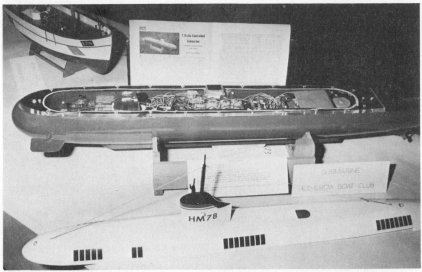

E Radio-controlled submarines exert a fascination for many modellers. This one was
   photographed at an exhibition.

Right, this French FSR-15 racer has clearly
been developed, witness the additional spray
strips riveted on. Radio stowage in
Tupperware bowl is good. Vertical plate is
race number.

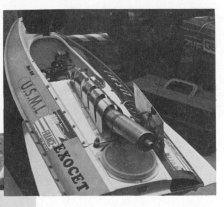

Left, central tuned pipe on
Scandinavian multi-racer – cover
on right includes sound insulation
for pipe. Radio hatch removal
entails undoing 10 nuts. Note single
bendable trim tab.

An F6 demonstration, actually a World War I naval battle with pyrotechnics etc. Even submarines get in on the act. Great fun.

Tugs manoeuvre an oil-drilling platform, a popular subject for F6 with platform fires or capture and recapture by opposing forces.

French battleship 'Richelieu' at a European Championship. Even with four screws, steering such a model precisely can be difficult, especially in a wind.

The bow of the 'Richelieu' shows the sort of detail in such models. Points would be lost for too glossy a hull finish and a little raggedness on the boot-top.

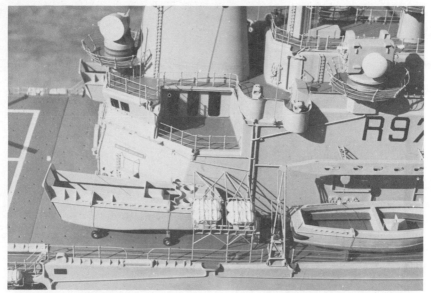

Nicely made scale model but the joint line for the removable part of the deck (at left) is rather prominent. Very difficult to hide such joints.

A lot of clutter on the deck such as on this oil-rig support vessel gives the opportunity of disguising flush hatch openings legitimately!